IFCoLog Journal of Logic and its Applications

Volume 2, Number 2

October 2015

Disclaimer

ISBN 978-1-84890-189-6
ISSN (E) 2055-3714
ISSN (P) 2055 3706

College Publications
Scientific Director: Dov Gabbay
Managing Director: Jane Spurr

http://www.collegepublications.co.uk

Printed by Lightning Source, Milton Keynes, UK

Editorial Board

Scope and Submissions

This journal considers submission in all areas of pure and applied logic, including:

- pure logical systems
- proof theory
- constructive logic
- categorical logic
- modal and temporal logic
- model theory
- recursion theory
- type theory
- nominal theory
- nonclassical logics
- nonmonotonic logic
- numerical and uncertainty reasoning
- logic and AI
- foundations of logic programming
- belief revision
- systems of knowledge and belief
- logics and semantics of programming
- specification and verification
- agent theory
- databases
- dynamic logic
- quantum logic
- algebraic logic
- logic and cognition
- probabilistic logic
- logic and networks
- neuro-logical systems
- complexity
- argumentation theory
- logic and computation
- logic and language
- logic engineering
- knowledge-based systems
- automated reasoning
- knowledge representation
- logic in hardware and VLSI
- natural language
- concurrent computation
- planning

This journal will also consider papers on the application of logic in other subject areas: philosophy, cognitive science, physics etc. provided they have some formal content.

Submissions should be sent to Jane Spurr (jane.spurr@kcl.ac.uk) as a pdf file, preferably compiled in LaTeX using the IFCoLog class file.

CONTENTS

ARTICLES

Gödel's Master Argument: What is it, and what can it do?

David Makinson
Department of Philosophy Logic and Scientific Method, London School of Economics

Abstract

This text is expository. We explain Gödel's 'Master Argument' for incompleteness as distinguished from the 'official' proof of his 1931 paper, highlight its attractions and limitations, and explain how some of the limitations may be transcended by putting it in a more abstract form that makes no reference to truth.

Keywords: Gödel, Master Argument, Incompleteness.

1 Introduction

Gödel's 'Master Argument' is sketched in his brief correspondence with Zermelo in late 1931. It is discussed in an influential 1984 article of Feferman [2], and may be found in books by several authors, most accessibly [8] and its website spin-off [9] . However, the argument is not as widely known as it should be, and its strengths and shortcomings compared to the 'official' proof appear not to have received much discussion. Moreover, an interesting abstraction on the Master Argument that overcomes some of the shortcomings, can be found only deep within the pages of specialist presentations such as [10] and [3], difficult to untangle from other material. The present article may thus be useful for those with limited time and energy but still wishing to have a proper understanding of what is going on in the Master Argument.

We begin by recalling the 1931 exchange of letters between Zermelo and Gödel, and itemize the background needed to continue reading. The Master Argument is then presented in its simplest available form, followed by a discussion balancing its attractions and limitations as well as an alleged philosophical weakness. We finally give a more abstract and powerful, but still easy, version of the Master Argument in which arbitrary 'oracles' take the place of 'truth in the intended model', thus transcending some of its limitations.

The author wishes to thank Jon Burton and Peter Smith for remarks on an ancestor of this text.

2 Autumn 1931

Gödel announced his incompleteness results in an abstract of 1930 and published them with proofs in his celebrated paper of 1931. Ernst Zermelo, already famous for his work on the axiom of choice and what we now call the Zermelo-Fraenkel axiomatization of set theory, read the 1931 paper and heard Gödel speak on it at a conference that summer. But he saw it as fatally flawed and ultimately not of great significance.

Both views appear to have stemmed from his disinterest in studying axiomatic systems that are formulated in finitary languages, *a fortiori* in doing so only by finitary means. Roughly speaking, Zermelo believed that we should be studying systems that embody broad swathes of mathematics, and that we should feel free to use any of the resources of mathematics in doing so. Both the formal systems studied and the reasoning used in that study could be infinitary along lines that he hoped, in the letters, to make precise at a later date.

This perspective evidently contrasts with that of Hilbert, which was adopted by Gödel in his published paper. The formal object-language that Gödel examines is defined by finite means and the investigation, conducted in a distinct and rather informal language, uses only finitary and constructive reasoning.

To be sure, in following decades logicians began relaxing these restrictions. Some investigated languages that are in one way or another infinitary, while others used free-wheeling methods with infinite sets, transfinite ordinals and the axiom of choice even when studying finitely generated systems. But in all cases they, like Hilbert and Gödel, continued to maintain a clear distinction between the system that is under study, formulated in an 'object-language', and the means used to study it, expressed in a (usually less formal) 'metalanguage'.

In contrast, Zermelo was unable or unwilling to make the distinction between object and metalanguage, and it seems to be that which led him to believe that Gödel's proofs harboured paradox. On 21 September 1931 he wrote to Gödel, hinting at his own general perspectives and outlining explicitly a contradiction that he claimed to have discovered in the paper.

Gödel replied on 12 October. He did not comment on the differences in general perspective, but responded in detail to the specific claim of paradox, carefully showing why his proof did not generate the contradiction that Zermelo thought he had found. At the same time, in an effort to help Zermelo see what was going on, he outlined the essence of his argument in a manner quite different from that of the version published earlier in the year. He did this again in an address in Princeton in 1934, but never elaborated it in print. After his death in 1978, the three letters constituting the Zermelo/Gödel exchange were found, published and translated. The proof there sketched came to be known as 'the Master Argument'.[1]

3 Background needed

We assume that the reader is familiar with the notation of first-order logic, and has seen the standard first-order axiomatization of the arithmetic of the natural numbers. We write *PA* (Peano Arithmetic) for the axiomatization, *LPA* for its formal language, N for the set of all natural numbers themselves.

On the semantic level, we presume familiarity with the notion of a model for a first-order theory, the recursive definition of satisfaction/truth in a model, and the concepts of soundness and completeness of a given theory with respect to a given model. On the syntactic level, the concepts whose definitions should already be familiar are those of a sentence (closed formula) of the language, free and bound variables, the consistency and negation-completeness of an arbitrary first-order theory and, for the particular case of *PA*, the notion of ω-consistency. With that basis, the reader will be able to verify from the definitions the easy parts of Figure 1 (all of them for *PA* and some for arbitrary first-order theories) namely the three vertical arrows and, given them, the following interrelations between the arrows:

- The diagonal full arrow follows from the top one,
- The diagonal dotted arrow follows from the diagonal full one,
- The bottom arrow follows from the diagonal dotted one,
- Conversely (and a little less obviously), the diagonal dotted arrow follows from the bottom one (the verification of this will be recalled in Section 4).

4 The Master Argument

The Master Argument has two parts: an Inexpressibility Lemma and an Expressibility Lemma; its conclusion arises from the collision of the two.

[1] Who coined the term 'Gödel's Master Argument'? The author has not been able to determine this with certainty. It is used as if familiar in [8], and already appeared tentatively in the first edition of that book (2007). In response an inquiry from the present author, Smith recalled that he had been using the phrase for some time in lectures in Cambridge, but could not remember whether he devised it himself or took it from another source. Of course, the term 'master argument' had already been used for certain other celebrated, although highly contested, demonstrations. It was applied in Greek antiquity to an argument of Diodorus Cronos about future and necessity, and since 1974 has been used to highlight one of Berkeley's arguments about existence and the mind (see the relevant Wikipedia articles).

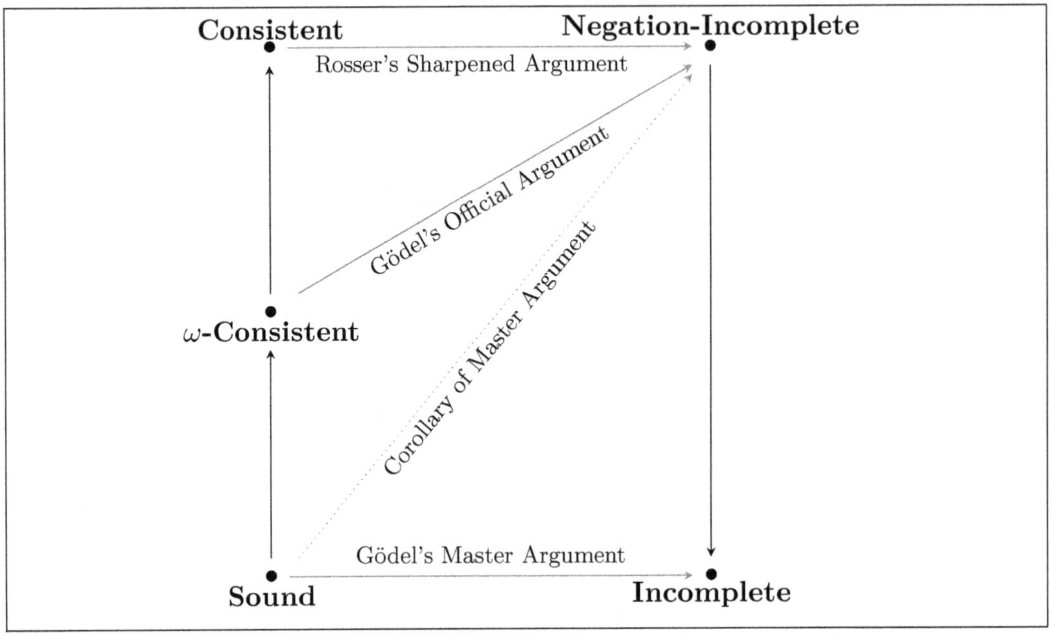

Figure 1: Gödel's first incompleteness theorem for *PA*

Definition 4.1. *A set $S \subseteq N$ is said to be* expressible *in LPA iff there is a formula $\varphi(x)$ of that language, with one free variable x, such that for all $n \in N$,*

$$n \in S \text{ iff } \varphi(\overline{n}) \text{ is true in the intended model for } PA$$

where $\overline{n}$ is the LPA numeral for n.

Fix any enumeration $\varphi_1, \varphi_2, \ldots, \varphi_i, \ldots, (i < \omega)$ of all the formulae in *LPA* whose sole free variable is x. Put D^+ to be the set of all natural numbers n such that $\varphi_n(\overline{n})$ is true in the intended model for *PA*, and let D^- be the set of all n such that $\varphi_n(\overline{n})$ is not true (i.e. false) in the same model. Clearly these two sets are complements of each other wrt. N, that is, $D^- = N \backslash D^+$ and $D^+ = N \backslash D^-$.

Lemma 4.2 (Inexpressibility Lemma[2]). *Neither D^- nor D^+ is expressible in the language of PA.*

[2]The formulation of the Inexpressibility Lemma 4.2 that is given here differs slightly from that sketched by Gödel in his letter to Zermelo, which has been followed in later presentations (e.g. Feferman, Smullyan, Fitting, Smith). On those accounts the lemma states the inexpressibility of truth itself (in other words, is exactly Tarski's Theorem), while on our account it states the inexpressibility of the sets D^+, D^-. Our formulation has the advantage that it simplifies the proof of the Inexpressibility Lemma 4.2, at the cost of then having to derive Tarski's Theorem from it

Proof. For D^-, suppose for reductio that it is expressible in *LPA*. Then by the definition of expressibility 4.1, there is a formula $\varphi(x)$ with x as sole free variable such that for all $n \in N, n \in D^-$ iff $\varphi(\overline{n})$ is true in the intended model. Now, $\varphi(x) = \varphi_k(x)$ for some $k \in N$. So, instantiating n to k we have: $k \in D^-$ iff $\varphi_k(\overline{k})$ is true in the intended model. But by the definition of D^-, we also have that $k \in D^-$ iff $\varphi_k(\overline{k})$ is not true in that model, giving a contradiction. Turning to D^+, it suffices to note that if D^+ is expressed by formula $\varphi(x)$ then D^- is expressed by $\neg\varphi(x)$. □

The second part of the Master Argument is a contrasting Expressibility Lemma. The sets D^+, D^- were defined using the notion of truth in the intended model of *PA*. We may also consider what happens if in the definitions we replace that notion by provability in the axiom system *PA*. Fix separate numberings of all formulae of *LPA* with just one free variable x, and of all derivations of *PA*. For brevity, write $|PA|$ for the set of all sentences that are theorems of *PA*. Put $D^+_{|PA|}$ to be the set of all natural numbers n such that $\varphi_n(\overline{n})$ is provable in *PA*, and let $D^-_{|PA|}$ be the set of all n such that $\varphi_n(\overline{n})$ is not provable in *PA*. Again these sets are complements of each other, so that one of them is expressible in the language of *PA* iff the other one is. But their behaviour is different from that of D^+, D^-. Indeed, as Gödel showed:

Lemma 4.3 (Expressibility Lemma). *If PA is sound wrt. its intended model, then both $D^+_{|PA|}$ and $D^-_{|PA|}$ are expressible in the language of PA.*

Proof. (sketch) It suffices to show this for $D^+_{|PA|}$. Consider the relation that holds between a derivation δ_m and a formula $\varphi_n(x)$ with just one free variable x iff the former is a derivation of $\varphi_n(\overline{n})$. Then (as outlined by Gödel with more detailed verifications in later presentations, e.g. [9]), assuming that the enumerations are in a certain technical sense 'acceptable', this relation is primitive recursive and so is *captured* in *PA* by some formulae $\psi(y, x)$ in the following sense: for all $m, n \in N$,

1. if δ_m stands in the relation to $\varphi_n(x)$ then $\psi(\overline{m}, \overline{n}) \in |PA|$ and
2. if δ_m does not stand in the relation to $\varphi_n(x)$ then $\neg\psi(\overline{m}, \overline{n}) \in |PA|$.

Now suppose that *PA* is sound wrt. its intended model. We want to show that the formula $\exists y\psi(y, x)$ expresses $D^+_{|PA|}$. That is, we need to check that for all $n \in N$,

$$n \in D^+_{|PA|} \text{ iff } \exists y\psi(y, \overline{n}) \text{ is true in the intended model.}$$

as is done in Section 5 point iii. The more common formulation eliminates any need for the latter derivation, but at the cost of a more complex proof of inexpressibility. This is a small matter of trade-offs.

Left to right: Suppose $n \in D^+_{|PA|}$. Then by definition, $\varphi_n(\overline{n}) \in |PA|$. Hence there is a derivation δ_m of $\varphi_n(\overline{n})$, so δ_m stands in the relation to $\varphi_n(x)$ so, by (1), $\psi(\overline{m}, \overline{n}) \in |PA|$, so by first-order logic $\exists y\psi(y, \overline{n}) \in |PA|$. Thus by the supposition of soundness, $\exists y\psi(y, \overline{n})$ is true in the intended model as desired.

Right to left: Suppose $n \notin D^+_{|PA|}$. Then by definition, $\varphi_n(\overline{n}) \notin |PA|$. Hence there is no derivation δ_m of $\varphi_n(\overline{n})$, so no δ_m stands in the relation to $\varphi_n(x)$ so, by (2), $\neg\psi(\overline{m}, \overline{n}) \in |PA|$ for all $m \in N$. Thus by soundness, $\neg\psi(\overline{m}, \overline{n})$ is true in the intended model for each $m \in N$, hence $\forall y\neg\psi(y, \overline{n})$ is true in the intended model, so $\exists y\psi(y, \overline{n})$ is not true in the intended model, as required. □

The semantic incompleteness of *PA* emerges immediately from the collision between these two Lemmas.

Theorem 4.4 (Gödel's First Incompleteness Theorem (semantic version)). *If PA is sound wrt. its intended model then it is incomplete (wrt. the same).*

Proof. It is immediate from the two results that the set of all sentences of *PA* that are *true* in the intended model is not the same as the set of all *provable* sentences of *PA*. Recall that soundness means that the latter set is included in the former; completeness is the converse. Thus if *PA* is sound, it is not complete. □

Corollary 4.5. *If PA is sound with respect to its intended model then it is negation-incomplete.*

Proof. This is simply an application to *PA* of the fact, mentioned in the last bullet point of Section 3, that for arbitrary first-order theories and intended models, the bottom arrow of the diagram implies the diagonal dotted one. Details: Suppose *PA* is sound wrt. its intended model. Then by incompleteness there is a sentence φ of *LPA* such that φ is true in the intended model of the theory but is not derivable in the theory. Hence $\neg\varphi$ is false in the intended model, so by soundness it is not derivable in the theory. Thus neither φ nor $\neg\varphi$ is derivable in the theory, which is to say that it is negation-incomplete. □

5 Three attractions

We articulate three important attractions of the Master Argument, before looking at shortcomings in the following section.

i. A striking feature of the Argument is the way that it decomposes the proof of the incompleteness theorem into two contrasting lemmas, thus providing a *simple overall architecture.* Moreover the diagonal argument for the first, 'negative' lemma is (in the present formulation) of the utmost simplicity, almost equal to that of Cantor's theorem in set theory. It is true that the second, 'positive' lemma remains long and tedious to check out in full detail, but at least the tedium is localized. Overall, one can say that the Master is the simplest and most transparent argument available for the semantic incompleteness, given soundness, of systems such as *PA*.

ii. Another feature of the proof is the prominent place that it gives to the notion of *expressibility* of subsets of N in the language *LPA* of Peano arithmetic, alongside the quite distinct notion of derivability of sentences from the axioms of *PA*. Since the formulation of the incompleteness theorem for *PA* speaks only of truth and provability, students easily assume that that is all it is about. Yet there are less expressive sub-languages of *PA* (indeed quite interesting ones) for which complete (and natural) axiomatizations are available. A well-known example is the sub-language in which zero, successor and addition remain but multiplication is absent.

   Expressive power and derivational power are thus two quite different capacities and can run in opposite directions. The former concerns the language alone and has nothing to do with provability from axioms; the latter is about which sentences, among those available in the language, are provable. The distinction is easily obscured by loose talk of the 'strength' of a system where it is left vague what kind of strength is meant. The salience that the Master Argument accords to the notion of expressibility has the merit of putting it on a par with that of provability.

iii. A third advantage of the Master Argument is that it reveals the close connection between Gödel's first incompleteness results and another celebrated theorem of mathematical logic: Tarski's 1933 theorem [11] on the indefinability of the notion of truth in *LPA* (or more expressive systems). We can state it (for *LPA*) as follows.

   Consider any enumeration $\psi_1, \psi_2, \ldots, \psi_i, \ldots, (i < \omega)$ of all sentences (formulae with no free variables) of *LPA*, and let T (the 'truth set') be the set of all $m \in N$ such that ψ_m is true in the intended model of *PA*. Then:

   Theorem 5.1 (Tarski's Theorem). *T is not expressible in LPA.*

   Proof. (sketch)

We apply the Inexpressibility Lemma 4.2. By definition, $n \in D^+$ iff $\varphi_n(\overline{n})$ is true in the intended model for *PA*, which holds iff there is an m with $\psi_m = \varphi_n(\overline{n})$ and $m \in T$. Now, it is not difficult (though tedious) to show that the relation that holds between n and m iff $\psi_m = \varphi_n(\overline{n})$, is expressible in *LPA*, so if T is also expressible in *LPA* then D^+ must also be so, contrary to the Inexpressibility Lemma 4.2. Thus T is not expressible in *LPA*. □

6 Two limitations

However, as presented above the Master Argument has two important limitations.

i. It yields only the semantic version of Gödel's first incompleteness theorem (soundness implies incompleteness and thus also negation-incompleteness). It does not give us Gödel's stronger syntactic version (ω-consistency implies negation-incompleteness); nor the yet stronger syntactic version obtained by Rosser (plain consistency implies negation-incompleteness); nor again Gödel's second incompleteness theorem (to the effect that if *PA* is consistent then its consistency cannot be proven by means executable within the system itself).

   However this ceiling on content can be raised. In the next section we formulate a more powerful version of the Inexpressibility Lemma 4.2, in which 'truth in the intended model' is replaced by an arbitrarily chosen 'oracle', and show how this, when combined with a corresponding variant of the Expressibility Lemma 4.3, gives us Gödel's syntactic version of the incompleteness result (the diagonal full arrow in the diagram) by a proof just as short and transparent as before.

   Nevertheless, it must be conceded that there is no visible way of using such oracles to obtain a similar proof of the Rosser version (top arrow), nor of Gödel's second theorem (not in the diagram), so there still remains a ceiling, albeit rather higher, on what the argument can achieve.

ii. This limitation is related to the fact that The Master Argument is *not constructive*. As Gödel put it in his letter of 12 October 1931, "it furnishes no construction of the undecidable statement". That is to say, it does not exhibit a specific sentence of *LPA* that is both true in the intended model and underivable from the axioms of *PA* (under the supposition of soundness). Nor does it give us a recipe for constructing such a sentence – it merely guarantees that there must be one. In contrast, the proof in the 1931 paper is constructive in every detail.

Assessment of this feature will vary with one's philosophy of mathematics. Doctrinal constructivists take the view that non-constructive proofs are invalid: they fail to show existence and at best can be taken as heuristic encouragements for devising properly constructive proofs of their results. There are few mathematicians and philosophers who would take such a position, but any who do must see the Master Argument as incorrect, with the real proof being the considerably more complex constructive one.

On the other hand, one may be constructively inclined without being doctrinal about it. There are results that most logicians are quite happy to establish non-constructively, for example the completeness of classical first-order (or even propositional) logic using the Lindenbaum-Henkin method of maxi-consistent sets of formulae. Since the 1930s mathematicians and logicians have gradually become more comfortable working with explicitly non-constructive principles, in particular the axiom of choice.

So a less radical view of the situation is that non-constructive proofs are perfectly valid, but give less information than their constructive counterparts when the latter are available. If one sees the additional information as important for one's purposes – as will often be the case in computer science – then one might describe oneself as a 'light constructivist'. While granting that non-constructive arguments are valid and frequently shorter and more transparent than constructive ones, the light constructivist is happy to put up with additional complexity in constructive reasoning to get further information.

In the present instance, the additional information that can be provided by a constructive proof appears to be needed for going on to Rosser's improved version of the completeness theorem, and likewise for Gödel's second incompleteness theorem. Its absence from the Master Argument is thus an important limitation. However, it may be said that the non-constructive proof still deserves a place *alongside* constructive ones because of the attractions mentioned above – simplicity and transparency of architecture, salient role of the notion of expressibility, and close connection to Tarski's theorem. In the author's opinion, it is the version to teach to non-specialists seeking a good understanding within a limited time-frame.

7 A philosophical weakness?

A further shortcoming, or at least potential one, was mentioned by Gödel in his letter to Zermelo: unlike the published proof, the Master Argument is "not intuitionistically unobjectionable".

Of course, non-constructivity is already objectionable to intuitionists, but the Master has a further feature that they do not accept: its use of the notions of *truth and falsehood* in mathematics. Intuitionists baulk at the idea that mathematical propositions have an objective truth-value beyond our ability to give intuitively satisfying demonstrations or refutations of them. For this reason, they do not accept in their full generality certain principles of classical propositional (and first-order) logic, most conspicuously the law of excluded middle, double negation elimination and one half of contraposition. But the very notion of expressibility, which features essentially in both lemmas for the Master Argument, is defined in terms of truth and falsehood in an intended model, and the law of excluded middle is implicit in e.g. the last sentence of the proof of the Inexpressibility Lemma 4.2.

In the early 1930s, intuitionism was a live option as a philosophy of mathematics, and its perspectives influenced the way in which Gödel presented his official proof of the incompleteness theorems. This is clear from a famous "crossed-out passage in an unsent reply" (Feferman's memorable phrase) written on 27 May 1970 to graduate student Yossef Balas. There Gödel said: "However in consequence of the philosophical prejudices of our time ... a concept of objective mathematical truth as opposed to demonstrability was viewed with greatest suspicion and widely rejected as meaningless." As Feferman observes in his paper of 1984, it is clear that when establishing his incompleteness result Gödel did not himself share that suspicion. But he nevertheless refrained from using the notions of mathematical truth and intended model out of an abundance of caution or, to put it more plainly, from fear of adverse reception by the mathematical community of the time.

Today intuitionistic logic is more an object of study than a code to live by. Few logicians and fewer mathematicians have any qualms about using the law of excluded middle or double negation elimination. Should we still retain any suspicions about the notion of *truth in the intended model* of a first-order theory? This is a philosophical question, and it would be rash to think that only one answer is possible. But many feel that there is no intrinsic difficulty with this concept. On the one hand, we can define the domain of the intended model, and the values to be given to the primitive operations of successor, addition and multiplication, within the confines of a quite small fragment of set theory; on the other hand we can define the truth-values of complex formulae, in that model, by recursion in the manner that was articulated by Tarski and is now standard.

On this view, there are really only two shortcomings to the Master Argument: a ceiling on what it shows and its non-constructivity. The canvassed philosophical weakness of relying on the notion of 'truth in the intended model of *PA*', is not a ground for serious concern.

Nevertheless, it is interesting to see that the Master Argument may be re-run on a purely syntactic plane without any reference to truth in the intended model. Thus, even if one has residual worries about that notion, they become irrelevant. The re-run has, moreover, a technical benefit: it can be done in such a way as to yield Gödel's syntactic version of the first incompleteness theorem (ω-consistency implies negation-incompleteness), thus raising somewhat the ceiling on content although remaining non-constructive. While a little more abstract than the basic version of the Master Argument, it is no more complex. We turn to it now.

8 The Master Argument without truth

We begin by generalizing the definition of expressibility. The definition in Section 4 took a set $S \subseteq N$ to be expressible in the language of *PA* iff there is a formula $\varphi(x)$ with one free variable x, such that for all $n \in N$, $n \in S$ iff $\varphi(\overline{n}) \in T$ where, we recall, T stands for the set of sentences that are true in the intended model for *PA*. Evidently, this definition continues to make sense if we replace T by another set of sentences of *LPA*; we can indeed generalize from T to an arbitrary set X of sentences, as follows.

Definition 8.1. *Let X be any set of sentences in the language of PA; we call it an 'oracle'. We say that that a set $S \subseteq N$ is* expressible according to (the oracle) X *iff there is a formula $\varphi(x)$ with one free variable x in LPA such that for all $n \in N$,*

$$n \in S \textit{ iff } \varphi(\overline{n}) \in X.$$

In particular, S is expressible according to the oracle T iff it is expressible *tout court*. The generalized definition 8.1 allows us to formulate a syntactic version of the Master Argument, using lemmas that follow the originals but with certain small changes.

As before, fix an enumeration of all formulae in the language of *PA* with just one free variable x. Generalize the definitions of D^- and D^+ thus: for any set X of sentences in the language of *PA*, put D^-_X (resp. D^+_X) to be the set of all natural numbers n such that $\varphi_n(\overline{n}) \notin X$ (resp. $\in X$). Clearly these two sets are complements of each other wrt. N, and as particular cases we have $D^-_T = D^-$ and $D^+_T = D^+$.

The Inexpressibility Lemma modulo an oracle 8.2 for D^-_X is formulated just as for D^-, but for D^+_X we need the hypothesis that X is well-behaved wrt. negation, in the sense that for every sentence φ in the language of *PA*, exactly one of $\varphi, \neg\varphi$ is in X. To appreciate the force of that hypothesis, note that one half of it (at least one of $\varphi, \neg\varphi$ is in X) is just negation-completeness, while the other half (at least one of $\varphi, \neg\varphi$ is not in X) is immediately implied by consistency. Indeed, if one assumes that X is closed under classical consequence, then the second half is equivalent to consistency. We have no need to make that assumption, but doing so would cause no harm to the argument.

Lemma 8.2 (Inexpressibility Lemma (modulo an oracle)). *Let X be any set of sentences in the language of PA. Then D^-_X is not expressible according to the oracle X. Moreover, if X is well-behaved wrt. negation, then D^+_X is not expressible according to X.*

Proof. For D^-_X, we argue exactly as before with X in place of T. Suppose for reductio that it is expressible according to X. Then by the definition of expressibility according to X, there is a formula $\varphi(x)$ with x as the sole free variable such that for all $n \in N$, $n \in D^-_X$ iff $\varphi(\overline{n}) \in X$. Now $\varphi(x) = \varphi_k(x)$ for some $k \in N$. So, instantiating n to k we have in particular $k \in D^-_X$ iff $\varphi_k(\overline{k}) \in X$. But by the definition of D^-_X we have $k \in D^-_X$ iff $\varphi_k(\overline{k}) \notin X$, giving a contradiction.

For D^+_X, suppose for *reductio* that it is expressible according to X by a formula $\varphi(x)$ and that X is well-behaved wrt. negation. Then for all $n \in N$, $n \in D^+_X$ iff $\varphi(\overline{n}) \in X$ iff $\neg\varphi(\overline{n}) \notin X$. But then $n \in D^-_X$ iff $n \notin D^+_X$ iff $\neg\varphi(\overline{n}) \in X$, so that D^-_X is expressed by the formula $\neg\varphi(x)$ according to X, contrary to what we have just shown. □

The Expressibility Lemma modulo an oracle 8.3 runs parallel to its unmodulated counterpart, with $D^+_{|PA|}$ replacing $D^+ = D^+_{|T|}$ and expressibility according to the oracle $|PA|$ replacing expressibility *tout court.* This forces two rejigs. Since truth is no longer involved, the lemma requires the condition of ω-consistency rather than soundness in the intended model; since expressibility of a set $S \subseteq N$ according to the oracle $|PA|$ does not in general follow immediately from the same for its complement $N \backslash S$, the lemma covers only $D^+_{|PA|}$ and not $D^-_{|PA|}$.

As before, we fix separate acceptable numberings of all formulae with just one free variable x and of all derivations of *PA*.

Lemma 8.3 (Expressibility Lemma (modulo an oracle)). *If PA is ω-consistent, then $D^+_{|PA|}$ is expressible according to the oracle $|PA|$.*

Proof. As in the proof of the Expressibility Lemma 4.3, consider the relation that holds between a derivation δ_m and a formula $\varphi_n(x)$ with just one free variable x iff the former is a derivation of $\varphi_n(\overline{n})$, and recall that this relation is captured in *PA* by some formulae $\psi(y, x)$ in the sense that for all $m, n \in N$,

1. if δ_m stands in the relation to $\varphi_n(x)$ then $\psi(\overline{m}, \overline{n}) \in |PA|$ and

2. if δ_m does not stand in the relation to $\varphi_n(x)$ then $\neg\psi(\overline{m},\overline{n}) \in |PA|$.

Now suppose that *PA* is ω-consistent. We want to show that the formula $\exists y\psi(y,x)$ expresses $D^+_{|PA|}$ according to the oracle $|PA|$; that is: for all $n \in N$, $n \in D^+_{|PA|}$ iff $\exists y\psi(y,\overline{n}) \in |PA|$.

Left to right: Suppose $n \in D^+_{|PA|}$. Then by definition, $\varphi_n(\overline{n}) \in |PA|$. Hence there is a derivation δ_m of $\varphi_n(\overline{n})$, so δ_m stands in the relation to $\varphi_n(x)$ so, by (1), $\psi(\overline{m},\overline{n}) \in |PA|$, so by classical logic $\exists y\psi(y,\overline{n}) \in |PA|$ as desired.

Right to left: Suppose $n \notin D^+_{|PA|}$. Then by definition, $\varphi_n(\overline{n}) \notin |PA|$. Hence there is no derivation δ_m of $\varphi_n(\overline{n})$, so no δ_m bears the relation to $\varphi_n(x)$ so, by (2), $\neg\psi(\overline{m},\overline{n}) \in |PA|$ for all $m \in N$. Thus by the ω-consistency of *PA*, $\exists y\psi(y,\overline{n}) \notin |PA|$ as desired. □

Those who relish fine detail may compare the verifications contained in the last two paragraphs with their counterparts in the Expressibility Lemma 4.3. Both directions have indeed become a little simpler as a result of dealing with the oracle $|PA|$ rather than the truth-set T: in the left to right direction, we could simply omit the last sentence of the previous version; in the converse direction, the last sentence brings ω-consistency into play in lieu of soundness modulo *PA*.

The syntactic version of Gödel's first incompleteness theorem appears immediately from the collision between the two oracular lemmas.

Theorem 8.4 (Gödel's First Incompleteness Theorem (syntactic version)). *If PA is ω-consistent then it is negation-incomplete.*

Proof. Suppose for *reductio* that *PA* is ω-consistent and negation-complete. Using ω-consistency, the Expressibility Lemma modulo an oracle 8.3 tells us that $D^+_{|PA|}$ is expressible according to $|PA|$. But ω-consistency immediately implies consistency so, combining that with negation-completeness, $|PA|$ is well-behaved wrt. negation. So the second part of the Inexpressibility Lemma modulo an oracle 8.2 tells us that $D^+_{|PA|}$ is not expressible according to $|PA|$, giving a contradiction. □

It is also possible to put Tarski's theorem on the undefinability of truth into a form that is no longer about truth in the intended model, but about an arbitrary oracle satisfying certain syntactic conditions. However, the details of both formulation and proof are a little more complex than we wish to handle here. We have gone only so far as is needed to render the Master Argument immune to the criticism that it uses the general notions of truth and falsehood of sentences of *PA*, and to raise the ceiling on its content to cover the syntactic version of Gödel's first incompleteness theorem. Readers who would like to see 'oracular' versions of Tarski's Theorem are directed to the texts of Smullyan and Fitting in the list of resources that follows.

References

[1] Hans-Dieter Ebbinghaus. *Ernst Zermelo: An Approach to his Life and Work.* Berlin: Springer, 2010. The correspondence with Gödel is discussed in section 4.10.

[2] Solomon Feferman. Kurt Gödel: conviction and caution. *Philosophia Naturalis*, 21:546–562, 1984. This influential paper was republished with minor additions as chapter 7 of the same author's book *In the Light of Logic*, Oxford: Oxford University Press 1998. Its theme is Gödel's outer caution about using the notion of truth in mathematics, as contrasted with his inner confidence in its meaningfulness.

[3] Melvin Fitting. *Incompleteness in the Land of Sets.* London: College Publications, 2007. An abstract form of the Master Argument is given in section 8.5. Also contains an abstract form of Tarski's Theorem.

[4] Kurt Gödel. Uber formal unentscheidbare Sätze der Principia Mathematica und Verwandter Systeme I. *Monatshefte für Mathematik und Physics*, 38:173–198, 1931. The 'official' version in its German original. A number of English translations are available, notably in Volume 1 of Gödel's collected works.

[5] Kurt Gödel. *Collected Works*, volume 1-5. Oxford: Clarendon Press, 1986-2003. The definitive, dual-language, collection. The letters are in volumes 5 and 6, ordered alphabetically by correspondent. Readers are urged to examine for themselves Zermelo's letter, Gödel's reply, and the response of Zermelo that ended the correspondence.

[6] Ivor Grattan-Guinness. In memoriam Kurt Gödel: his 1931 correspondence with Zermelo on his incompletability theorem. *Historia Mathematica*, 6:294–304, 1979. The first publication (in German) of Gödel's reply to Zermelo, dated 12 October 1931, and the latter's response of 29 October. The initial letter of Zermelo, of 21 September, was found later by John Dawson in Gödel's Nachlass and first published (in German) in his note 'Completing the Gödel-Zermelo correspondence' *Historia Mathematica* 12 (1985): 66-70.

[7] Roman Murawski. Undefinability of truth. the problem of priority: Tarski vs Gödel. *History and Philosophy of Logic*, 19:153–160, 1998. Discusses the historical relationship between Gödel's work on incompleteness and Tarski's work on the undefinability of truth.

[8] Peter Smith. *An Introduction to Gödel's Theorems.* Cambridge: Cambridge University Press, 2nd edition, 2013. The Master Argument is sketched in section 27.5, but without abstraction to a version without truth.

[9] Peter Smith. *Gödel without (too many) tears.* version of 20 February 2015. The text is based on Smith 2013, but is more selective, concise and lively. It is perhaps the most readable among accounts that go deeply into the machinery of Gödel's incompleteness theorems. The Master Argument is in a box in section 50.

[10] Raymond Smullyan. *Diagonalization and Self-Reference.* New York: Oxford University Press, 1994. Seeks maximum generality; not to be tackled lightly. Abstract versions of the Master Argument appear several times, beginning with Theorem 2.2. Also contains an abstract version of Tarski's Theorem.

[11] Alfred Tarski. *Pojęcie prawdy w językach nauk dedukcynjnych*. Nakladem Towarzysta Naukowego Warszawskiego: Warsaw. An English translation, 'The concept of truth in formalized languages', may be found in pp. 152-278 of the same author's collection *Logic, Semantics Metamathematics*, Oxford: Clarendon Press.

Received March 2015

Cut-Free Proof Systems for Geach Logics

Melvin Fitting
melvinfitting.org
melvin.fitting@gmail.edu

Abstract

Prefixed tableaus for modal logics have been around since the early 1970s, and are quite familiar by now. Rather recently it was found that they were dual to nested sequents, which have a complicated history but which also trace back to the 1970's. Both have provided very natural proof systems for the most common modal logics, including those in the so-called modal cube. In this paper we add some simple machinery to both prefixed tableaus and to nested sequents, producing cut-free proof systems for all logics axiomatized by Geach formulas, that is, by axiom schemes of the form $\Diamond^k\Box^l X \supset \Box^m\Diamond^n X$. This again provides proof mechanisms for the modal cube, but mechanisms of a different nature than usual. But further, it provides proof mechanisms for an infinite family of modal logics, and does so in a modular way with a clear separation between logical and structural rules. The version of nested sequents presented here has a direct relationship with the formal machinery of [23], and can be thought of as a notational variant of a natural and interesting fragment of what can be handled using that methodology.

1 Introduction

Over the years, sequent calculi and tableau systems for modal logics have evolved into many forms. In appearance some have been elegant, some perhaps a bit ornate. Some formalisms turned out to be embeddable in others, sometimes unexpectedly. In this paper we add yet more species to the genus: tableau and sequent calculi specifically for logics in the *Geach family*, also known as the *Scott-Lemmon logics*.

Definition 1.1 (Geach family). *A Geach formula scheme is of the form $\Diamond^k\Box^l X \supset \Box^m\Diamond^n X$, where $k, l, m, n \geq 0$. (These are also known in the literature as Lemmon-Scott axioms.) Following [6], this scheme is denoted $G^{k,l,m,n}$. Scheme $G^{k,l,m,n}$ is axiomatically sound and complete (indeed, canonical) with respect to frames meeting*

Thank an anonymous referee for the comments that led to section 11.

the condition: if $w_1\mathcal{R}^k w_2$ *and* $w_1\mathcal{R}^m w_3$ *then for some* w_4, $w_2\mathcal{R}^l w_4$ *and* $w_3\mathcal{R}^n w_4$, *a kind of diamond property. We overload the notation* $\mathsf{G}^{k,l,m,n}$ *to mean: a Geach formula scheme, or the corresponding semantic condition, or the corresponding tableau rule (Section 3), or the corresponding nested sequent rule (Section 8). Context can adequately sort things out. We systematically use* **Geach** *for a* set *of* $\mathsf{G}^{k,l,m,n}$, *as formulas, or semantic conditions, or tableau rules, or nested sequent rules, and we call* **Geach** *a* Geach logic.

Many of the most common modal logics, T, K4, S4, S5, among others, are Geach logics. There are infinitely many different logics in the Geach family. We give new versions of *prefixed tableaus* and of *nested sequents* appropriate for all Geach logics, though at the cost of additional machinery added to the common versions. Our systems make a clear distinction between logical rules and structural rules. Logical rules are common to all the modal logics we treat; differences are reflected entirely in structural rules.

In [4] Brünnler proposed "structural modal rules," and conjectured the existence of modular systems using these rules. An attempt to prove the conjecture was made in [5] but there was a flaw in the argument; the conjecture is not true in general. In [20] the error was fixed and a truly modular version was created, capable of handling not only all 15 logics in the classical S5 cube, but of providing distinct modular systems corresponding to distinct, though equivalent, axiomatizations. The machinery in the present paper also does this using different, but related, machinery. It goes beyond the earlier work in that it handles the entire Geach family, which is infinite.

We note that [17] already studies nested sequent rules for a restricted class of Geach logics, $\mathsf{G}^{k,l,m,n}$ where either $l = 1$ and $n = 0$ or else $l = 0$ and $n = 1$. This in turn is based on display calculi for Geach logics (among others), in [19]. Also [20] provides a nested sequent calculus for the S5 modal cube. All these examine issues of constructive cut elimination and other important properties of proofs that we do not consider here. The advantages of the systems given here are these: simple and natural additions to the usual prefixed tableau or nested sequent machinery provide proof systems covering all Geach logics in a uniform and (relatively) uncomplicated way.

A few words about how things evolved may not be inappropriate, though this is not meant to be a proper history. (Some details can be found in [10].) Axiomatic proofs display nothing but formulas. Gentzen added the comma (with a context dependent interpretation) and the arrow, obtaining proof systems for classical and intuitionistic logic that provided special insights. The earliest sequent calculi for modal logics followed Gentzen's format. Several common modal logics can be handled this way, but even S5 is a problem. The earliest tableau systems were analogous, and can be thought of as dual to sequent calculi.

As time went on, additional machinery was added to both tableau and sequent calculi. This machinery fell into two categories. On the one hand, the language itself might be enlarged. If this route is taken, any additional machinery is governed directly by logical rules, proof schemes, axioms, etc. *Hybrid* logics take this route. Nominals are added to the language, naming possible worlds. With nominals available, a much wider variety of logics have natural tableau and sequent proof systems, [2]. Perhaps the farthest reaching work along language expansion lines is somewhat intermediate, enlarging the language used in proofs while primary interest centers in formulas not involving the enhanced machinery, [23]. Possible worlds and an accessibility relation in effect become primitives of a formal language. The resulting general machinery, *labelled sequent calculi*, will be related to the present work in Section 11. In a different direction, *prefixed tableaus*, [8, 9, 16, 21], and more generally *labeled deductive systems*, [14], add machinery to a basic sequent or tableau calculus, but like Gentzen's arrow and comma this machinery is not part of the formula language itself. The additional machinery is not manipulated as part of an expanded logic, but syntactically and from the outside of the language, so to speak. We follow this second path here, and we should say a little about our motivation.

Justification logics originated with an *explicit* counterpart of the modal logic S4, [1]. A justification logic contains explicit *justification terms* instead of a modal operator. These justification terms reflect the underlying reasoning and logical dependence inherent in a modal theorem. Each theorem of S4 has a provable counterpart in the justification logic LP, a connection called a *Realization Theorem*. The range of modal logics having justification counterparts gradually grew beyond S4 and closely related logics. Recently, in unpublished work, I found that all logics in the Geach family have justification counterparts—an infinite family. My proof of Realization for these logics is non-constructive. Constructive proofs of Realization always make use of so-called *cut free* proof systems. The real point is, proof systems are needed that have the subformula property and preserve subformula polarity. Gentzen style sequents have been used to prove constructive Realization results, and so have Smullyan style tableaus, hypersequents, prefixed tableaus, and nested sequents. So far, at least, proof methods that make use of expanded languages have not been used successfully for Realization proofs. It is plausible that if we could produce a cut-free proof system for Geach logics without expanding the underlying language, it might be possible to give a constructive Realization proof uniformly for all Geach logics. This is the hope, and the present paper is intended to be a step in this direction.

2 Prefixed Tableaus

It will be useful to sketch standard prefixed tableaus before moving on to our modifications. There are some minor novelties in our presentation, and these need some mention as well.

It is convenient for us to make use of signed formulas and uniform notation. We begin by briefly sketching what all this means. Formulas are built up from propositional letters using propositional connectives $\wedge$, $\vee$, $\supset$, $\neg$, and modal operators $\Box$ and $\Diamond$. (It is useful to allow $\top$ to appear in tableau proofs, and we do so. It serves as a place-holder and will not appear as part of a more complex formula. When we come to nested sequents, the empty sequent plays a similar role.) A *signed formula* is $T\,X$ or $F\,X$, where X is a formula. (Unsigned formulas could have been used as well. It makes no essential difference. Using signed formulas makes certain things easier.) Signed formulas involving a binary connective are grouped into categories, those that behave conjunctively (α) and those that behave disjunctively (β). Similarly signed modal formulas are grouped into those behaving like necessity (ν) and those behaving like possibility (π). This is what is referred to as *uniform notation.* Details are given in the following tables, which also define *components* for each of these signed formulas.

α	α_1	α_2
$T\,X \wedge Y$	$T\,X$	$T\,Y$
$F\,X \vee Y$	$F\,X$	$F\,Y$
$F\,X \supset Y$	$T\,X$	$F\,Y$

β	β_1	β_2
$F\,X \wedge Y$	$F\,X$	$F\,Y$
$T\,X \vee Y$	$T\,X$	$T\,Y$
$T\,X \supset Y$	$F\,X$	$T\,Y$

ν	ν_0
$T\,\Box X$	$T\,X$
$F\,\Diamond X$	$F\,X$

π	π_0
$F\,\Box X$	$F\,X$
$T\,\Diamond X$	$T\,X$

Informally, one reads $T\,X$ as asserting that X is true (in some particular circumstance), and $F\,X$ as asserting that X is false. Each α is then equivalent to the informal conjunction of α_1 and α_2, while β cases act disjunctively. Similarly a ν signed formula informally holds at a possible world if ν_0 holds at all accessible worlds, while a π signed formula holds if π_0 holds at some accessible world.

Definition 2.1 (Path Sequences). A path sequence *is a finite sequence of positive integers, such as* 1.3.2.5. *A period is customarily used as a separator. A path sequence can be empty; denoted by* ϵ. A prefixed, *signed formula is an expression of the form* $\sigma\,Z$, *where* σ *is a path sequence and* Z *is a signed formula. Throughout, we systematically use* σ *to represent an arbitrary path sequence and* Z *to represent an arbitrary signed formula.*

Prefixed signed (or unsigned) formulas are used in prefixed tableau systems for many common modal logics. Such proof systems have been around for many years and are described in a number of sources—[10] for instance. Commonly the empty path sequence is not allowed and path sequences start with 1, but these are minor points. Nothing basic is affected by our small modifications, but making them now fits better with the new machinery to be introduced shortly. Informally, a prefix names a possible world, and $\sigma\, Z$ says that Z is so in the world named by σ. The structure of sequences is supposed to provide a syntactic representation of accessibility: informally $\sigma.n$ names a world accessible from σ. For the example 1.3.2.5 mentioned in Definition 2.1, ϵ names an arbitrary world, 1 names a successor of this world (2 would name a different one), 1.3 names a successor of the world named by 1, and so on.

Finally we introduce some special terminology and machinery for working with prefixed tableaus.

Definition 2.2 (Path Sequence Terminology). *If σ is a path sequence and n is a positive integer, by $\sigma.n$ we mean the result of adding n to the end of path sequence σ. We will also need $n.\sigma$, which is the result of adding n to the beginning of σ.*

Prefixed tableaus have prefixed signed formulas at tableau nodes. We say a path sequence σ occurs *on a tableau branch if $\sigma\, Z$ is on the branch for some signed formula Z.*

A positive integer n is new *in a tableau if it does not appear in any path sequence occurring on any branch in the tableau. A path sequence σ is* not new *on a tableau branch if $\sigma\, Z$ occurs on the branch for some Z.*

Remark There are some differences in terminology between items in the definition above and what is common in the literature. Usually one requires newness of a *prefix*, and not just of an integer, in certain tableau rules. Our requirements are stronger (but imply the usual ones). The stronger version makes it simpler to describe a translation from our new kind of tableaus into our new kind of nested sequents, but it makes no real difference for tableau soundness or completeness. Also, note that our condition for not newness is still for entire prefixes.

As an example, we now describe a prefixed tableau system for the weakest normal modal logic, K. A prefixed tableau proof of a modal formula X begins with the single prefixed signed formula $\epsilon\, F\, X$, where ϵ is the empty path sequence. Informally, beginning with $\epsilon\, F\, X$ amounts to supposing X could be false at some world, corresponding to ϵ, in some model. Tableau expansion rules are then applied until a closed tableau, which is formally a tree, is produced (or not). A closed tableau informally means the supposition that X could be false somewhere is impossible. Here are the rules and conditions for doing this.

Prefixed Tableau Closure A tableau branch is *closed* if it contains $\sigma\, T\, X$ and $\sigma\, F\, X$, for some formula X and some path sequence σ. If X is atomic, the branch is *atomically closed.* If all branches of a tableau are (atomically) closed, the tableau is (atomically) closed.

Next we have rules for expanding a tableau, stated using uniform notation.

Classical Prefixed Tableau Rules

$$\frac{\sigma\,\alpha}{\begin{array}{c}\sigma\,\alpha_1\\ \sigma\,\alpha_2\end{array}} \qquad \frac{\sigma\,\beta}{\sigma\,\beta_1 \mid \sigma\,\beta_2} \qquad \frac{\sigma\, T\, \neg X}{\sigma\, F\, X} \qquad \frac{\sigma\, F\, \neg X}{\sigma\, T\, X}$$

Modal Prefixed Tableau Rules

$$\frac{\sigma\,\nu}{\sigma.n\,\nu_0} \qquad\qquad \frac{\sigma\,\pi}{\sigma.n\,\pi_0}$$

where $\sigma.n$ is not new on the branch (for ν); where n is new in the tableau (for π).

Example 2.3. *Here is a very simple prefixed* K *tableau proof, of* $\Box(P \supset Q) \supset (\Box P \supset \Box Q)$.

$$\begin{array}{ll}
\epsilon\, F\, \Box(P \supset Q) \supset (\Box P \supset \Box Q) & 1.\\
\epsilon\, T\, \Box(P \supset Q) & 2.\\
\epsilon\, F\, \Box P \supset \Box Q & 3.\\
\epsilon\, T\, \Box P & 4.\\
\epsilon\, F\, \Box Q & 5.\\
1\, F\, Q & 6.\\
1\, T\, P & 7.\\
1\, T\, P \supset Q & 8.
\end{array}$$

$$1\, F\, P \quad 9. \qquad\qquad 1\, T\, Q \quad 10.$$

In this, 2 and 3 are from 1 by Classical α*, as are 4 and 5 from 3; 6 is from 5 by* Modal π *(integer 1 is new on the branch at this point, and we write* $\epsilon.1$ *simply as* 1*); 7 is from 4 and 8 is from 2 by* Modal ν *(prefix 1 is not new on the branch now); 9 and 10 are from 8 by* Classical β*. Closure is by 7 and 9, and by 6 and 10.*

Single-use conditions can be imposed. All rules except Modal ν can be restricted to a single application to a given prefixed signed formula on a branch. Likewise *atomic closure* conditions can be imposed. All closure of branches must be atomic. Soundness and completeness can be proved with these conditions in place.

Several other modal logics can be captured by adding additional rules to those above. For instance, adding the following yields K4.

$$\frac{\sigma\,\nu}{\sigma.n\,\nu}$$

where $\sigma.n$ is
not new on the branch

Our new machinery will provide quite different rules for logics like K4, so we do not pursue this further here.

3 Set Prefixed Tableaus

A key point about tableaus as described in Section 2 is that these are appropriate for logics for which tree models suffice. In a tree there is a unique path from the origin to each node, so it is really ambiguous whether prefixes designate paths or nodes—it doesn't matter. Frames that are trees are not sufficient for Geach formulas generally. For instance, $\Diamond\Box X \supset \Box\Diamond X$ is complete for frames having the confluence property: if $w_1\mathcal{R}w_2$ and $w_1\mathcal{R}w_3$ then for some w_4, $w_2\mathcal{R}w_4$ and $w_3\mathcal{R}w_4$. This involves two paths from w_1 to w_4, and so we are not dealing with a frame having a tree structure.

We now specifically want to think of a path sequence as representing a *path* in a frame, and not a node. With this understanding it is natural to start with the empty sequence, and we have done so. Integers represent edges.

If integer sequences represent paths, we still need a representation for the possible worlds of frames. We take these to be *sets of sequences*. Then, for example, the set $\{1.2.3, 1.4\}$ intuitively represents a node in a frame such that there are (at least) two paths leading to it, starting from some arbitrary node represented by the set containing the empty sequence $\{\epsilon\}$. Using this path/set of paths representation, we introduce tableau systems of a new kind, for all Geach logics. In particular, this provides new tableau systems for some familiar logics, T, K4, S4, S5 among them, and also for S4.2, and an infinite family of other logics as well. We call these *set prefixed tableaus.*

There are connections between prefixed tableaus and tableaus for hybrid logic. The present work is related to that presented in [3]. As noted earlier, the essential difference is that the hybrid approach is uniform across a wide range of logics, but it requires expansion of the usual modal language, while the set prefixed approach works with the original language, though it expands the machinery allowed in formal proofs. There may also be useful relationships with [22], but this remains to be examined.

Definition 3.1 (Path Sets). *A* path set *is a finite, non-empty set of path sequences (Definition 2.1). A* set prefixed signed formula *is* $\Sigma\, Z$ *where* Σ *is a path set and* Z *is a signed formula. We systematically use* Σ *to represent an arbitrary path set.*

It is useful to think of $\Sigma\, Z$ as shorthand for $\{\sigma\, Z \mid \sigma \in \Sigma\}$. This motivates some of the terminology that follows.

Definition 3.2 (Path set terminology). *If* Σ *is a path set and* n *is a positive integer, by* $\Sigma.n$ *we mean the result of adding* n *to the end of every path sequence in* Σ. *That is,* $\Sigma.n = \{\sigma.n \mid \sigma \in \Sigma\}$.

Set prefixed tableaus will have set prefixed signed formulas at nodes. We say a path set Σ occurs on a tableau branch *if* $\Sigma\, Z$ *is on the branch for some signed formula* Z. *We say a path sequence* σ occurs on a branch *if* $\sigma \in \Sigma$ *where* Σ *is a path set that occurs on the branch. If* $\Sigma\, Z$ *occurs on a tableau branch and* $\sigma \in \Sigma$, *we say* $\sigma\, Z$ occurs on the branch *too.*

A positive integer n *is* new *on a tableau branch if it does not appear in any path sequence occurring on the branch. It is new in the tableau if it is new on every branch. When working with a particular tableau branch, we write* $\Sigma_1 \to \Sigma_2$ *if both path sets* Σ_1 *and* Σ_2 *occur on the branch and* $\sigma \in \Sigma_1$, $\sigma.n \in \Sigma_2$ *for some* σ *and* n.

Note that, speaking very informally, $\Sigma_1 \to \Sigma_2$ tells us there is an edge from the node named by Σ_1 to the node named by Σ_2.

We now present tableau rules that make use of set prefixed signed formulas. A set prefixed tableau proof of a modal formula X begins with the single set prefixed signed formula $\{\epsilon\}\, F\, X$, where ϵ is the empty path sequence. Beginning with $\{\epsilon\}\, F\, X$ informally amounts to supposing X could be false at some world, corresponding to $\{\epsilon\}$, in some model. Tableau expansion rules are then applied. As usual, a closed tableau informally means the supposition that X could be false somewhere is impossible. Here are the rules and conditions for doing this. Not surprisingly, they resemble those from Section 2, at least for a while.

Set Prefixed Closure A tableau branch is *closed* if it contains both $\Sigma\, T\, X$ and $\Sigma\, F\, X$, for some formula X and some path set Σ. If X is atomic, the branch is *atomically closed.* If all branches of a tableau are (atomically) closed, the tableau is (atomically) closed.

Next we have versions of the usual propositional and modal rules, once again stated using uniform notation.

Classical Set Prefixed Tableau Rules

$$\frac{\Sigma\,\alpha}{\begin{array}{c}\Sigma\,\alpha_1\\ \Sigma\,\alpha_2\end{array}} \qquad \frac{\Sigma\,\beta}{\Sigma\,\beta_1 \mid \Sigma\,\beta_2} \qquad \frac{\Sigma\,T\,\neg X}{\Sigma\,F\,X} \qquad \frac{\Sigma\,F\,\neg X}{\Sigma\,T\,X}$$

Classical Set Prefixed Modal Rules

$$\frac{\Sigma_1\,\nu}{\begin{array}{c}\Sigma_2\,\nu_0\\ \text{where } \Sigma_1 \to \Sigma_2\end{array}} \qquad \frac{\Sigma\,\pi}{\begin{array}{c}\Sigma.n\,\pi_0\\ \text{where } n \text{ is}\\ \text{new in the tableau}\end{array}}$$

As far as we have gone in rule presentation, a tableau construction beginning with $\{\epsilon\}\,F\,X$ can only involve set prefixed signed formulas with singleton sets as prefixes. Machinery for something more complex has not yet been introduced. If we temporarily identify $\{\sigma\}$ with σ, at this point we have the usual prefixed tableau rules for the modal logic K, from Section 2, in disguise. We move on to rules that deal specifically with the internal structure of path sets, and it is now that more complex sets enter the picture.

If two path sets appear on a branch and some path sequence is common to both, it must be that the node of the frame that the two path sets informally designate is the same, since both allow getting to the node by the same path and paths lead to unique nodes. In this case the two path sets can be merged.

Set Prefix Union

$$\frac{\Sigma_1\,Z}{\begin{array}{c}(\Sigma_1 \cup \Sigma_2)\,Z\\ \text{provided } \Sigma_2 \text{ on branch}\\ \text{and } \Sigma_1 \cap \Sigma_2 \neq \emptyset\end{array}}$$

If edge n leads from one node in a frame to another, *any* path leading to the first node continues to the second via n. This motivates the following rule, in which the premise is that $\sigma.n\,Z$ occurs on a tableau branch, according to Definition 3.2.

Set Prefix Continuation

$$\frac{\sigma.n\,Z}{\begin{array}{c}\Sigma.n\,Z\\ \text{provided } \Sigma \text{ on branch}\\ \text{and } \sigma \in \Sigma\end{array}}$$

We need some special notation before introducing the Geach rule (or properly, rule scheme). Note that $\top$ is now allowed to appear on a tableau branch. It serves as a place-holder—we can't have a set prefix with no signed formula that it prefixes.

Notation and Terminology
$\mathbb{P}$ is the set of positive integers, and $\mathbb{P}^n = \{\langle k_1, \ldots, k_n\rangle \mid k_i \in \mathbb{P}\}$. We call each k_i a *component* of $\vec{t} = \langle k_1, \ldots, k_n\rangle$.

We say $\vec{t} \in \mathbb{P}^n$ is *new* in a tableau if each component is new and no two components are the same. If also $\vec{u} \in \mathbb{P}^m$, we say $\vec{t}$ and $\vec{u}$ *don't overlap* if they do not share a component.

Suppose $\vec{t} \in \mathbb{P}^n$, say $\vec{t} = \langle k_1, \ldots, k_n\rangle$. For a path sequence σ, we write $\sigma.\vec{t}$ as short for $\sigma.k_1.k_2.\cdots.k_n$. For a path set Σ, we write $\Sigma.\vec{t}$ as short for $\{\sigma.\vec{t} \mid \sigma \in \Sigma\}$. And finally, we write $(\Sigma \oplus \mathsf{init}(\vec{t}))\, T \top$ as short for the sequence:

$$\begin{array}{l} \Sigma.k_1\, T\top \\ \Sigma.k_1.k_2\, T\top \\ \vdots \\ \Sigma.k_1.k_2.\cdots.k_{n-1}\, T\top \end{array}$$

Now, here is our central new tableau construct. Note that it is entirely about manipulation of set prefixes—signed formulas play no essential role.

Set Prefixed Geach Rule Scheme for $\mathsf{G}^{k,l,m,n}$

$$\frac{\begin{array}{c}\sigma.\vec{t} \in \Sigma_1 \text{ on branch}, \vec{t} \in \mathbb{P}^k \\ \sigma.\vec{u} \in \Sigma_2 \text{ on branch}, \vec{u} \in \mathbb{P}^m\end{array}}{\begin{array}{c}(\Sigma_1 \oplus \mathsf{init}(\vec{v}))\, T\top \\ (\Sigma_2 \oplus \mathsf{init}(\vec{w}))\, T\top \\ (\Sigma_1.\vec{v} \cup \Sigma_2.\vec{w})\, T\top \\ \text{where} \\ \vec{v} \in \mathbb{P}^l \text{ is new} \\ \vec{w} \in \mathbb{P}^n \text{ is new} \\ \text{and } \vec{v} \text{ and } \vec{w} \text{ don't overlap}\end{array}}$$

This rule scheme mimics the semantic conditions for $\mathsf{G}^{k,l,m,n}$ as given in Definition 1.1. Informally, the premises that $\sigma.\vec{t}$ and $\sigma.\vec{u}$ are on a branch say that from a single possible world (reachable via path σ) one can follow paths of lengths k and m. The semantic conditions tell us that from these positions things can be brought back together via continuation paths of lengths l and n. These paths, except for the final common node, are represented by $(\Sigma_1 \oplus \mathsf{init}(\vec{v}))\, T\top$ and $(\Sigma_2 \oplus \mathsf{init}(\vec{w}))\, T\top$, and they are brought together at $(\Sigma_1.\vec{v} \cup \Sigma_2.\vec{w})\, T\, \top$, completing the diamond. The newness and non-overlapping conditions are analogous to the newness condition of the usual existential instantiation rule. They play a central role in the soundness argument in Section 5.

In Section 2 we noted that certain *single-use* conditions could be imposed on prefixed tableaus. The same is true for set prefixed tableaus. The Classical Rules and the Modal π rule can be restricted to a single application to any particular set prefixed signed formula on a branch. Our completeness proof works with this restriction imposed. It also works with closure required to be atomic. This completes the presentation of the tableau rules. Examples are in the next section.

4 Set Prefixed Tableau Examples

We give several examples of Geach style tableau rules for common modal logics. It should be noted that rules for transitivity, symmetry, and so on are quite different from those familiar with standard prefixed tableau systems.

Example 4.1. ***Geach Rule for $G^{0,1,2,0}$***
Axiomatically, $G^{0,1,2,0}$ is the familiar scheme $\Box X \supset \Box\Box X$. The corresponding Geach Rule is the following.

$$\frac{\begin{array}{c}\sigma \in \Sigma_1 \textit{ on branch}\\ \sigma.a.b \in \Sigma_2 \textit{ on branch}\end{array}}{\begin{array}{c}(\Sigma_1.c \cup \Sigma_2)\, T\, \top\\ \textit{where } c \textit{ is new}\end{array}}$$

Incidentally, the dual version of $\Box X \supset \Box\Box X$ is $\Diamond\Diamond X \supset \Diamond X$ and it is also a Geach scheme, $G^{2,0,0,1}$. It leads to the same rule as above, but with the roles of Σ_1 and Σ_2 switched around. We will not mention this duality phenomenon when discussing additional examples below. We continue by showing this rule in use.

Proof of $\Box\Diamond\Diamond P \supset \Box\Diamond P$ in K4

$$
\begin{array}{lll}
\{\epsilon\} & F\,\Box\Diamond\Diamond P \supset \Box\Diamond P & 1. \\
\{\epsilon\} & T\,\Box\Diamond\Diamond P & 2. \\
\{\epsilon\} & F\,\Box\Diamond P & 3. \\
\{1\} & F\,\Diamond P & 4. \\
\{1\} & T\,\Diamond\Diamond P & 5. \\
\{1.2\} & T\,\Diamond P & 6. \\
\{1.2.3\} & T\,P & 7. \\
\{1.4, 1.2.3\} & T\,\top & 8. \\
\{1.4, 1.2.3\} & F\,P & 9. \\
\{1.4, 1.2.3\} & T\,P & 10.
\end{array}
$$

The reasons are as follows. Recall that ϵ is the empty path sequence. 2 and 3 are from 1 by Classical α*; 4 is from 3 by* Modal π*; 5 is from 2 by* Modal ν*; 6 is from 5 and 7 is from 6 by* Modal π*; 8 is from 4 (or 5), and 7, by* $\mathsf{G}^{0,1,2,0}$*; 9 is from 4 by* Modal ν*; and 10 is from 7 and 8 (or 9) by* Prefix Union*. Closure is from 9 and 10, and happens to be atomic.*

Example 4.2. ***Geach Rule for*** $\mathsf{G}^{0,1,0,0}$
The Geach axiom scheme is, of course, $\Box X \supset X$. The corresponding Geach Rule is the following.

$$
\frac{\begin{array}{c}\sigma \in \Sigma_1 \text{ on branch} \\ \sigma \in \Sigma_2 \text{ on branch}\end{array}}{\begin{array}{c}(\Sigma_1.a \cup \Sigma_2)\,T\,\top \\ \text{where } a \text{ is new}\end{array}}
$$

In this rule the premises say that Σ_1 and Σ_2 overlap. Since we have the Prefix Union Rule*, without loss of generality we can simply take the sets to be the same.*

Derived Geach Rule for $\mathsf{G}^{0,1,0,0}$

$$
\frac{\Sigma \text{ on branch}}{\begin{array}{c}(\Sigma.a \cup \Sigma)\,T\,\top \\ \text{where } a \text{ is new}\end{array}}
$$

Example 4.3. ***Geach Rule for*** $\mathsf{G}^{1,1,0,0}$
The Geach axiom scheme is $\Diamond\Box X \supset X$, and the Geach rule is the following.

$$
\frac{\begin{array}{c}\sigma.a \in \Sigma_1 \text{ on branch} \\ \sigma \in \Sigma_2 \text{ on branch}\end{array}}{\begin{array}{c}(\Sigma_1.b \cup \Sigma_2)\,T\,\top \\ \text{where } b \text{ is new}\end{array}}
$$

By combining $G^{0,1,2,0}$ and $G^{0,1,0,0}$ we have a tableau system for S4. It can be extended to one for S5 using either $\Diamond\Box X \supset \Box X$ or $\Diamond\Box X \supset X$. We have given the Geach Rule for the second of these. The first version then becomes provable.

Proof of $\Diamond\Box P \supset \Box P$ using $\mathsf{G}^{1,1,0,0}$ and $\mathsf{G}^{0,1,2,0}$

$$
\begin{array}{lll}
\{\epsilon\} & F\,\Diamond\Box P \supset \Box P & 1. \\
\{\epsilon\} & T\,\Diamond\Box P & 2. \\
\{\epsilon\} & F\,\Box P & 3. \\
\{1\} & T\,\Box P & 4. \\
\{1.2, \epsilon\} & T\,\top & 5. \\
\{3\} & F\,P & 6. \\
\{1.2.3, 3\} & F\,P & 7. \\
\{1.4, 1.2.3, 3\} & T\,\top & 8. \\
\{1.4, 1.2.3, 3\} & T\,P & 9. \\
\{1.4, 1.2.3, 3\} & F\,P & 10.
\end{array}
$$

In this 2 and 3 are from 1 by Classical α; 4 is from 2 by Modal π; 5 is from 3 and 4 by Geach Rule $G^{1,1,0,0}$ where $\Sigma_1 = \{1\}$, $\Sigma_2 = \{\epsilon\}$ $\sigma = \epsilon$, $a = 1$, and $b = 2$; 6 is from 3 by Modal π; 7 is from 5 and 6 by Prefix Continuation where $\Sigma = \{1.2, \epsilon\}$, $\sigma = \epsilon$, and $a = 3$; 8 is from 4 and 7 by Geach Rule $G^{0,1,2,0}$ where $\Sigma_1 = \{1\}$, $\Sigma_2 = \{1.2.3, 3\}$, $\sigma = 1$, $a = 2$, $b = 3$, and $c = 4$; 9 is from 4 and 8 by Modal ν; 10 is from 7 and 9 by Prefix Union where $\Sigma_1 = \{1.2.3, 3\}$ and $\Sigma_2 = \{1.4, 1.2.3, 3\}$; closure is by 9 and 10.

Example 4.4. ***Geach Rule for $G^{1,1,1,1}$***
Axiomatically $G^{1,1,1,1}$ is the scheme $\Diamond\Box X \supset \Box\Diamond X$. It is the archetypical Geach scheme. When combined with S4 we have the logic S4.2. The rule is the following.

$$
\frac{\begin{array}{c}\sigma.a \in \Sigma_1 \textit{ on branch} \\ \sigma.b \in \Sigma_2 \textit{ on branch}\end{array}}{(\Sigma_1.c \cup \Sigma_2.d)\,T\,\top}
$$
$$
\begin{array}{c}\textit{where} \\ c \textit{ and } d \textit{ are new and distinct}\end{array}
$$

Proof of $(\Diamond\Box P \wedge \Diamond\Box Q) \supset \Diamond(P \wedge Q)$ using $\mathsf{G}^{1,1,1,1}$ and $\mathsf{G}^{0,1,2,0}$

$\{\epsilon\}\quad F\,(\Diamond\Box P \wedge \Diamond\Box Q) \supset \Diamond(P \wedge Q)\quad 1.$
$\{\epsilon\}\quad T\,\Diamond\Box P \wedge \Diamond\Box Q\quad 2.$
$\{\epsilon\}\quad F\,\Diamond(P \wedge Q)\quad 3.$
$\{\epsilon\}\quad T\,\Diamond\Box P\quad 4.$
$\{\epsilon\}\quad T\,\Diamond\Box Q\quad 5.$
$\{1\}\quad T\,\Box P\quad 6.$
$\{2\}\quad T\,\Box Q\quad 7.$
$\{1.3, 2.4\}\quad T\,\top\quad 8.$
$\{5, 1.3, 2.4\}\,T\,\top\quad 9.$
$\{5, 1.3, 2.4\}\,T\,P\quad 10.$
$\{5, 1.3, 2.4\}\,T\,Q\quad 11.$
$\{5, 1.3, 2.4\}\,F\,P \wedge Q\quad 12.$

$\{5,\ 1.3,\ 2.4\}\,F\,P\quad 13.\qquad \{5,\ 1.3,\ 2.4\}\,F\,Q\quad 14.$

Here 2 and 3 are from 1 by Classical α*, as are 4 and 5 from 2; 6 is from 4 and 7 from 5 by* Modal π*; 8 is from 6 and 7 by* $\mathsf{G}^{1,1,1,1}$*; 9 is from 1 and 8 by* $\mathsf{G}^{0,1,2,0}$*; 10, 11, and 12 are from 6, 7, and 3 by* Modal ν*; 13 and 14 are from 12 by* Classical β.

Example 4.5. ***Geach Rule for*** $\mathsf{G}^{1,2,1,2}$ *The Geach axiom scheme is* $\Diamond\Box\Box X \supset \Box\Diamond\Diamond X$*. The rule is as follows.*

$$\frac{\begin{array}{c}\sigma.a \in \Sigma_1 \text{ on branch} \\ \sigma.b \in \Sigma_2 \text{ on branch}\end{array}}{\begin{array}{c}\Sigma_1.c\,T\,\top \\ \Sigma_2.e\,T\,\top \\ (\Sigma_1.c.d \cup \Sigma_2.e.f)\,T\,\top \\ \text{where } c, d, e, f \text{ are} \\ \text{new and distinct}\end{array}}$$

Proof of $(\Diamond\Box\Box P \wedge \Diamond\Box\Box Q) \supset \Diamond\Diamond\Diamond(P \wedge Q)$ using $\mathsf{G}^{1,2,1,2}$

$$
\begin{array}{ll}
\{\epsilon\}\, F\, (\Diamond\Box\Box P \wedge \Diamond\Box\Box Q) \supset \Diamond\Diamond\Diamond(P \wedge Q) & 1. \\
\{\epsilon\}\, T\, \Diamond\Box\Box P \wedge \Diamond\Box\Box Q & 2. \\
\{\epsilon\}\, F\, \Diamond\Diamond\Diamond(P \wedge Q) & 3. \\
\{\epsilon\}\, T\, \Diamond\Box\Box P & 4. \\
\{\epsilon\}\, T\, \Diamond\Box\Box Q & 5. \\
\{1\}\, T\, \Box\Box P & 6. \\
\{2\}\, T\, \Box\Box Q & 7. \\
\{1\}\, F\, \Diamond\Diamond(P \wedge Q) & 8. \\
\{1.3\}\, T\, \top & 9. \\
\{2.5\}\, T\, \top & 10. \\
\{1.3.4, 2.5.6\}\, T\, \top & 11. \\
\{1.3\}\, T\, \Box P & 12. \\
\{2.5\}\, T\, \Box Q & 13. \\
\{1.3\}\, F\, \Diamond(P \wedge Q) & 14. \\
\{1.3.4, 2.5.6\}\, T\, P & 15. \\
\{1.3.4, 2.5.6\}\, T\, Q & 16. \\
\{1.3.4, 2.5.6\}\, F\, P \wedge Q & 17.
\end{array}
$$

$$
\{1.3.4,\ 2.5.6\}\, F\, P \quad 18. \qquad \{1.3.4,\ 2.5.6\}\, F\, Q \quad 19.
$$

2 and 3 are from 1 by Classical α*, as are 4 and 5 from 2; 6 is from 4 and 7 is from 5 by* Modal π*; 8 is from 3 by* Modal ν*; 9, 10, and 11 are from 6 and 7 by* $\mathsf{G}^{1,2,1,2}$*; 12, 13, and 14 are from 6, 7, and 8 by* Modal ν*; 15, 16, and 17 are from 12, 13, and 14 also by* Modal ν*; 18 and 19 are from 17 by* Classical β*.*

5 Tableau Soundness

We show soundness for a set prefixed tableau system containing a set Geach of Geach Rules. Throughout this section $\mathcal{M} = \langle \mathcal{G}, \mathcal{R}, \mathcal{V} \rangle$ is a Kripke model with possible worlds $\mathcal{G}$, accessibility relation $\mathcal{R}$, and valuation $\mathcal{V}$. We assume the frame meets the semantic conditions for the members of Geach. We write $\mathcal{M}, w \Vdash X$ to indicate that formula X is true at world w of model $\mathcal{M}$. We extend notation to signed formulas, writing $\mathcal{M}, w \Vdash T\,X$ if $\mathcal{M}, w \Vdash X$, and $\mathcal{M}, w \Vdash F\,X$ if $\mathcal{M}, w \nVdash X$. We often choose a particular possible world w_0 and call it the *actual world*. The choice of an actual world is arbitrary, and the terminology is for convenience.

Definition 5.1 (Satisfiability). *The frame of Kripke model $\mathcal{M}$ is a directed graph. Let θ be a mapping from a set P of positive integers to edges of this directed graph. We call θ an* edge mapping. *Assume a possible world w_0 has been designated as actual world. Further, suppose $\mathcal{B}$ is a branch of a set prefixed tableau. We say θ* satisfies $\mathcal{B}$ in $\mathcal{M}$ with respect to w_0 *if the following conditions are met.*

1. *The domain P of θ consists of those positive integers that appear in path sequences on branch $\mathcal{B}$.*
2. *If $\sigma = n_0.n_1.n_2.\ldots.n_k$ is a path sequence appearing on $\mathcal{B}$ then $\theta(n_0)$, $\theta(n_1)$, $\theta(n_2)$, ..., $\theta(n_k)$ is a path in $\langle\mathcal{G},\mathcal{R}\rangle$ beginning at w_0. We denote this path by $\theta(\sigma)$ and its terminal node (possible world) by $\mathcal{T}(\theta(\sigma))$. Then $\theta(\sigma)$ is a path in $\langle\mathcal{G},\mathcal{R}\rangle$ from the designated actual world to $\mathcal{T}(\theta(\sigma))$.*
3. *If Σ is a path set that occurs on branch $\mathcal{B}$ then $\mathcal{T}(\theta(\sigma))$ is the same for all path sequences σ appearing in Σ. We refer to this common node as $\mathcal{T}(\theta(\Sigma))$.*
4. *If $\Sigma\, Z$ occurs on branch $\mathcal{B}$, where Z is a signed formula, then $\mathcal{M}, \mathcal{T}(\theta(\Sigma)) \Vdash Z$.*

We say a branch $\mathcal{B}$ of a set prefixed tableau is satisfiable *if there is some Kripke model $\mathcal{M}$, some choice of actual world w_0 of $\mathcal{M}$, and some edge assignment θ so that θ satisfies $\mathcal{B}$ in $\mathcal{M}$ with respect to w_0. A tableau is* satisfiable *if some branch is satisfiable.*

Theorem 5.2. *Suppose we have a set prefixed tableau, allowing rules from the set* Geach, *and it is satisfiable in a model $\mathcal{M}$ meeting the semantic conditions for members of* Geach. *If a tableau rule is applied, the resulting tableau is again satisfiable in $\mathcal{M}$.*

Proof. The proof of this is along the lines usual with modal tableaus. Suppose we have a satisfiable tableau and a tableau rule is applied on branch $\mathcal{B}$ of it. If a branch of the tableau other than $\mathcal{B}$ was satisfiable, it still is after a rule application on $\mathcal{B}$—this is a trivial case. Now suppose it is $\mathcal{B}$ that is satisfiable. There are many cases to be checked, one for each tableau rule. We check three cases and leave the rest to the reader. For each case assume tableau branch $\mathcal{B}$ is satisfiable in model $\mathcal{M}$ using edge mapping θ with respect to actual world w_0, and $\mathcal{B}$ has a rule applied on it.

Set Prefix Union Rule. Suppose $\Sigma_1\, Z$ is on $\mathcal{B}$, Σ_2 occurs on $\mathcal{B}$, $\Sigma_1 \cap \Sigma_2 \neq \emptyset$, and we add $(\Sigma_1 \cup \Sigma_2)\, Z$ to $\mathcal{B}$. The set of positive integers appearing in path sequences on $\mathcal{B}$ has not changed, so θ still meets condition 1 from Definition 5.1 for the extension of $\mathcal{B}$. Similarly for condition 2. Since $\mathcal{T}(\theta(\tau))$ is the same for all path sequences τ in Σ_1, and similarly for Σ_2, and since some path sequence appears in both, then $\mathcal{T}(\theta(\tau))$ is the same for all τ in $\Sigma_1 \cup \Sigma_2$; hence condition 3. Since $\Sigma_1\, Z$ is on $\mathcal{B}$, by the induction hypothesis $\mathcal{M}, \mathcal{T}(\theta(\Sigma_1)) \Vdash Z$. Then $\mathcal{M}, \mathcal{T}(\theta(\Sigma_1 \cup \Sigma_2)) \Vdash Z$, using condition 3, and hence we have condition 4.

Set Prefix Continuation Rule. Suppose path set Σ occurs on $\mathcal{B}$, $\sigma \in \Sigma$, $\sigma.a\, Z$ is on $\mathcal{B}$, and we add $\Sigma.a\, Z$ to $\mathcal{B}$. The conditions on θ ensure that edge $\theta(a)$ goes from $\mathcal{T}(\theta(\Sigma))$ to $\mathcal{T}(\theta(\Sigma.a))$, and this is enough to confirm condition 3 for the extended branch. The other conditions are straightforward.

Geach Rule for $\mathsf{G}^{k,l,m,n}$. Assume $\mathcal{M}$ meets the frame conditions corresponding to $\mathsf{G}^{k,l,m,n}$, and an application of tableau Geach Rule $\mathsf{G}^{k,l,m,n}$ is made, extending $\mathcal{B}$. More specifically, suppose path sets Σ_1 and Σ_2 occur on $\mathcal{B}$, $\sigma.\vec{t} \in \Sigma_1$, $\sigma.\vec{u} \in \Sigma_2$ ($\vec{t} \in \mathbb{P}^k$, $\vec{u} \in \mathbb{P}^m$), and we add the members of $(\Sigma_1 \oplus \mathsf{init}(\vec{v}))\, T\, \top$, the members of $(\Sigma_2 \oplus \mathsf{init}(\vec{w}))\, T\, \top$, and $(\Sigma_1.\vec{v} \cup \Sigma_2.\vec{w})\, T\, \top$ to $\mathcal{B}$ ($\vec{v} \in \mathbb{P}^l$, $\vec{w} \in \mathbb{P}^n$, both new, and not overlapping). We must show the resulting branch is satisfiable.

We know that before $\mathcal{B}$ was extended, θ satisfied $\mathcal{B}$. Then there is a path, $\theta(\vec{t})$, of length k from $\mathcal{T}(\theta(\sigma))$ to $\mathcal{T}(\theta(\sigma.\vec{t})) = \mathcal{T}(\theta(\Sigma_1))$, and a path, $\theta(\vec{u})$, of length m from $\mathcal{T}(\theta(\sigma))$ to $\mathcal{T}(\theta(\sigma.\vec{u})) = \mathcal{T}(\theta(\Sigma_2))$. By the semantic condition for $\mathsf{G}^{k,l,m,n}$ there is a possible world, call it $w \in \mathcal{G}$, with a path from $\mathcal{T}(\theta(\Sigma_1))$ to w of length l, and a path from $\mathcal{T}(\theta(\Sigma_2))$ to w of length n. We extend θ to a mapping θ' by defining it on the positive integers that occur in $\vec{v}$ and $\vec{w}$. Since these are new and don't overlap, we can do this freely. Let θ' assign to the successive positive integers in $\vec{v}$ the edges on the path from $\mathcal{T}(\theta(\Sigma_1))$ to w. Likewise let θ' assign to the positive integers in $\vec{w}$ the edges on the path from $\mathcal{T}(\theta(\Sigma_2))$ to w. It is now straightforward to check that the branch extending $\mathcal{B}$ is satisfiable using θ'. We omit the details.

□

Theorem 5.3 (Soundness). *If formula X is provable using the general set prefixed tableau rules and a set* Geach *of Geach Rules, all from Section 3, then X is true at every possible world of every Kripke model meeting the semantic conditions for members of* Geach.

Proof. Suppose X is provable, but is false in model $\mathcal{M} = \langle \mathcal{G}, \mathcal{R}, \mathcal{V} \rangle$ at possible world w_0, where the tableau proof uses members of Geach and $\mathcal{M}$ meets the semantic conditions for Geach. We derive a contradiction.

Use w_0 as the actual world. The initial tableau has a single branch, containing a single entry, $\{\epsilon\}\, F\, X$, and this branch is satisfiable in $\mathcal{M}$ using for edge assignment the mapping with empty domain. Then by Theorem 5.1, all subsequent steps in the tableau construction are satisfiable. But since X is provable, a closed tableau can be reached. This is a contradiction since a closed tableau cannot be satisfiable. □

6 Tableau Completeness

Throughout this section we assume we are using the set prefixed tableau rules allowing a set Geach of the Geach Rule schemes. We will show completeness with respect to Kripke models whose frames meet the semantic conditions for Geach.

It is possible to specify a systematic and fair tableau construction procedure (with a single-use restriction and requiring atomic closure). Fairness simply means that any applicable rule is eventually applied. We omit details of such a construction procedure. Now suppose we follow a fair construction procedure, and it does not produce a closed tableau. This tableau must have an open branch. (Of course the construction may not terminate, in which case an infinite tableau will be generated. Such a tableau will have an infinite branch by König's Lemma, and this branch will be open, or we would have stopped applying rules to it at a finite stage.) Completeness is a consequence of the following.

Theorem 6.1. *Let $\mathcal{B}$ be an open branch of a set prefixed tableau, using rules from Geach, where the tableau has been constructed using a fair procedure (following a single-use restriction and requiring atomic closure). The collection of set prefixed, signed formulas on $\mathcal{B}$ is a satisfiable set, satisfiable in a Kripke model meeting the semantic conditions for Geach.*

Proof. We use $\mathcal{B}$ to construct a model $\mathcal{M}$. Details of the construction are introduced and facts are established one item at a time. Justification of facts follows their statement, when necessary.

1. Call two path sequences σ_1 and σ_2 *equivalent*, written $\sigma_1 \sim \sigma_2$, if there is some path set Σ that occurs on branch $\mathcal{B}$ with both as members. This is an equivalence relation on the set of path sequences. We denote the equivalence class containing path sequence σ by $[\sigma]$.

In verifying that we have an equivalence relation, reflexivity and symmetry are trivial. For transitivity, suppose $\sigma_1 \sim \sigma_2$ and $\sigma_2 \sim \sigma_3$; say $\sigma_1, \sigma_2 \in \Sigma_1$, and $\sigma_2, \sigma_3 \in \Sigma_2$, where both Σ_1 and Σ_2 occur on $\mathcal{B}$. Since σ_2 belongs to both, $\Sigma_1 \cap \Sigma_2 \neq \emptyset$, so using the Set Prefix Union Rule (and the fairness of the tableau construction), $\Sigma_1 \cup \Sigma_2$ also occurs on $\mathcal{B}$. It contains σ_1 and σ_3, so $\sigma_1 \sim \sigma_3$.

2. The set $\mathcal{G}$ of possible worlds of the model $\mathcal{M}$ that we are constructing is the set of equivalence classes of path sequences, using the equivalence relation $\sim$.

3. If $\sigma_1 \sim \sigma_2$ and $\sigma_1.a$ occurs on $\mathcal{B}$ then $\sigma_1.a \sim \sigma_2.a$.

We verify item 3. Assume $\sigma_1 \sim \sigma_2$; say both σ_1 and σ_2 are in path set Σ, on $\mathcal{B}$. Also assume $\sigma_1.a$ occurs on $\mathcal{B}$, say in the prefixed signed formula $\sigma_1.a\, Z$. By the Set Prefix Continuation Rule and the fairness of the tableau construction, $\Sigma.a\, Z$ also occurs on $\mathcal{B}$. Since both σ_1 and σ_2 are in Σ, both $\sigma_1.a$ and $\sigma_2.a$ will be in $\Sigma.a$, hence $\sigma_1.a \sim \sigma_2.a$.

4. An accessibility relation $\mathcal{R}$ is defined on $\mathcal{G}$ as follows. $[\sigma_1]\mathcal{R}[\sigma_2]$ if $\sigma_1.a \in [\sigma_2]$ for some positive integer a. An equivalent simpler version is: $[\sigma]\mathcal{R}[\sigma.a]$, provided $\sigma.a$ appears on $\mathcal{B}$.

Note that item 4 does not depend on the particular representative chosen for $[\sigma_1]$ because of 3. There is an important consequence of this, which will be needed below in item 9. Suppose $[\sigma_1]\mathcal{R}[\sigma_2]\mathcal{R}[\sigma_3]$. Then $\sigma_1.a.b \in [\sigma_3]$ for some a and b, by the following argument. Since $[\sigma_1]\mathcal{R}[\sigma_2]$, then $\sigma_1.a \in [\sigma_2]$ for some a. Since $[\sigma_2]\mathcal{R}[\sigma_3]$, $\sigma_2.b \in [\sigma_3]$ for some b. But $\sigma_1.a \sim \sigma_2$, so by 3, $\sigma_1.a.b \sim \sigma_2.b$, and hence $\sigma_1.a.b \in [\sigma_3]$. Of course this can be continued to longer chains of accessibility.

5. If $\sigma_1 \sim \sigma_2$, and $\sigma_1\, Z$ occurs on branch $\mathcal{B}$ for signed formula Z, so does $\sigma_2\, Z$.

We verify item 5. Assume $\sigma_1 \sim \sigma_2$; say $\sigma_1, \sigma_2 \in \Sigma_1$, where Σ_1 occurs on $\mathcal{B}$. Also assume $\sigma_1\, Z$ occurs on $\mathcal{B}$; say $\sigma_1 \in \Sigma_2$ where $\Sigma_2\, Z$ occurs on $\mathcal{B}$. Then Σ_1 and Σ_2 overlap, on σ_1, so by the Prefix Union Rule, $(\Sigma_1 \cup \Sigma_2)\, Z$ is on $\mathcal{B}$. Since $\sigma_2 \in \Sigma_1$, $\sigma_2\, Z$ is on $\mathcal{B}$.

More can be obtained by a similar argument.

6. Suppose $\sigma\, T\, X$ and $\sigma\, F\, X$ occur on $\mathcal{B}$. Then $\mathcal{B}$ is closed. More generally, we have closure if $\sigma_1\, T\, X$ and $\sigma_2\, F\, X$ are on $\mathcal{B}$, where $\sigma_1 \sim \sigma_2$.

We verify the first part of 6; the second part then follows using 5. Assume $\sigma\, T\, X$ and $\sigma\, F\, X$ occur on $\mathcal{B}$; say $\sigma \in \Sigma_1$ where $\Sigma_1\, T\, X$ is on $\mathcal{B}$ and $\sigma \in \Sigma_2$ where $\Sigma_2\, F\, X$ is on $\mathcal{B}$. Since Σ_1 and Σ_2 overlap, using the Prefix Union Rule $(\Sigma_1 \cup \Sigma_2)\, T\, X$ and $(\Sigma_1 \cup \Sigma_2)\, F\, X$ are on $\mathcal{B}$, and this satisfies the closure definition.

7. The valuation $\mathcal{V}$ of our model is defined as follows. For a propositional letter P, $\mathcal{V}(P) = \{[\sigma] \mid \sigma\, T\, P \text{ occurs on } \mathcal{B}\}$. (Item 5 is relevant here.) We have finished construction of our model, $\mathcal{M} = \langle \mathcal{G}, \mathcal{R}, \mathcal{V} \rangle$.

8. Truth Lemma. *If* $\sigma\, Z$ *occurs on* $\mathcal{B}$ *then* $\mathcal{M}, [\sigma] \Vdash Z$.

The proof of item 8 is by induction on the complexity of signed formula Z.

We begin with the atomic case. If $Z = T\, P$, the result is by valuation definition. Next we show that $\sigma\, F\, P$ on $\mathcal{B}$ and $\mathcal{M}, [\sigma] \Vdash P$ together lead to a contradiction. Well, suppose we had both. By the second item $[\sigma] \in \mathcal{V}(P)$ so by 7 $\sigma\, T\, P$ occurs on $\mathcal{B}$, but then by 6, $\mathcal{B}$ would be closed. Notice that only atomic closure of the tableau construction is needed here.

For the induction steps, assume the result is known for signed formulas simpler than Z. There are several cases. The propositional connective cases are virtually the same as in every tableau completeness proof. We leave these cases to the reader. We cover the modal cases involving $\Box$ in more detail. Those involving $\Diamond$ are similar and are omitted.

Suppose $\sigma\, T\, \Box X$ occurs on $\mathcal{B}$. A possible world in $\mathcal{G}$ accessible from $[\sigma]$ must be of the form $[\sigma.a]$ for some positive integer a where $\sigma.a$ occurs on $\mathcal{B}$. Say $\sigma \in \Sigma_1$ where $\Sigma_1\, T\, \Box X$ is on $\mathcal{B}$, and $\sigma.a \in \Sigma_2$, which is also on $\mathcal{B}$. By Modal Rule ν, $\Sigma_2\, T\, X$ must be on $\mathcal{B}$ (since $\Sigma_1 \to \Sigma_2$), so $\sigma.a\, T\, X$ is on $\mathcal{B}$. By the induction hypothesis, $\mathcal{M}, [\sigma.a] \Vdash X$. And since a was arbitrary, we have this for *any* accessible world, and hence $\mathcal{M}, [\sigma] \Vdash \Box X$.

Finally, suppose $\sigma\, F\, \Box X$ occurs on $\mathcal{B}$, say $\sigma \in \Sigma$ and $\Sigma\, F\, \Box X$ is on $\mathcal{B}$. By Modal Rule π (even single-use), $\Sigma.n\, F\, X$ is on $\mathcal{B}$ for some n. Then $\sigma.n\, F\, X$ is on $\mathcal{B}$. $[\sigma.n] \in \mathcal{G}$ and $[\sigma]\mathcal{R}[\sigma.n]$. By the induction hypothesis, $\mathcal{M}, [\sigma.n] \not\Vdash X$, and so $\mathcal{M}, [\sigma] \not\Vdash \Box X$.

This completes the argument for 8.

9. The frame of model $\mathcal{M} = \langle \mathcal{G}, \mathcal{R}, \mathcal{V} \rangle$ satisfies the semantic conditions of Geach.

We verify 9. Assume we use Geach Rule $\mathsf{G}^{k,l,m,n}$ in the tableau construction. We show the corresponding semantic condition from Definition 1.1 holds in $\mathcal{M}$.

Suppose $[\sigma_1], [\sigma_2], [\sigma_3] \in \mathcal{G}$ and $[\sigma_1]\mathcal{R}^k[\sigma_2]$ and $[\sigma_1]\mathcal{R}^m[\sigma_3]$. By repeated use of the definition of $\mathcal{R}$ from 4 (and the comments immediately following that definition) there must be some $\vec{t} \in \mathbb{P}^k$ so that $\sigma_1.\vec{t} \in [\sigma_2]$ and hence $[\sigma_2] = [\sigma_1.\vec{t}]$. Similarly there is some $\vec{u} \in \mathbb{P}^m$ so that $[\sigma_3] = [\sigma_1.\vec{u}]$. Then there are path sets Σ_1 and Σ_2 on $\mathcal{B}$ with $\sigma_1.\vec{t} \in \Sigma_1$ and $\sigma_1.\vec{u} \in \Sigma_2$. By the Geach Rule for $\mathsf{G}^{k,l,m,n}$ there are $\vec{v} \in \mathbb{P}^l$ and $\vec{w} \in \mathbb{P}^n$ so that $(\Sigma_1.\vec{v} \cup \Sigma_2.\vec{w})\, T\, \top$ is on $\mathcal{B}$. Then $\sigma_1.\vec{t}.\vec{v} \sim \sigma_1.\vec{u}.\vec{w}$, so $[\sigma_1.\vec{t}.\vec{v}] = [\sigma_1.\vec{u}.\vec{w}] \in \mathcal{G}$. Finally, it is easy to see that $[\sigma_1.\vec{t}]\mathcal{R}^l[\sigma_1.\vec{t}.\vec{v}]$ and $[\sigma_1.\vec{u}]\mathcal{R}^n[\sigma_1.\vec{u}.\vec{w}]$.

Now all the pieces of completeness are in place. If X is not provable, there is no closed tableau starting with $\{\epsilon\}\, F\, X$. A fair tableau construction will produce a tableau with an open branch, $\mathcal{B}$, to which the items above can be applied. A model $\mathcal{M} = \langle \mathcal{G}, \mathcal{R}, \mathcal{V} \rangle$ will be produced, and by 9 it will satisfy the Geach semantic conditions. In this, $[\epsilon]$ will be a possible world and, since $\{\epsilon\}\, F\, X$ is on $\mathcal{B}$, by 8, $\mathcal{M}, [\epsilon] \not\Vdash X$, and thus X is not valid. □

7 Nested Sequent Systems

Tableau systems and sequent calculi generally go together. Roughly speaking, one is the other upside-down. Smullyan style tableaus and Gentzen style sequent calculi correspond in this way, and this extends to destructive modal tableaus and sequent calculi. In [11] it was shown that prefixed modal tableaus and nested sequents correspond. In [18] it was further shown that nested sequents correspond to the subclass of labelled sequents called labelled tree sequents. There are certain ideas that keep recurring in various forms—an argument for naturalness. Starting in Section 8 we present *indexed* nested sequents, which (mostly) correspond to set prefixed modal tableaus in a similar way. But first, in this section we sketch the basics of ordinary nested sequents, presenting a system just for modal K.

Nested sequents are sequent calculi that allow certain kinds of *deep reasoning.* The machinery to manage this consists of allowing sequents to appear inside sequents, which can appear inside sequents, etc. Rules apply at any level of nesting. All sequents are one-sided here. One sided sequents can be defined using sequences, multi-sets, or sets. To keep things simple, we have opted for a set-based approach. Sometimes the empty sequent is disallowed. Here it is allowed. Also we use signed formulas, while formulas without signs are commonly used in the literature on nested sequents.

Definition 7.1. *A* nested sequent *is a finite set of signed formulas and nested sequents.*

There are standard conventions for writing nested sequents, and we follow them. At the top level all enclosing curly brackets are omitted. At deeper levels, a nested sequent that is a member of another nested sequent is represented by listing its members in square brackets, and is called a *boxed sequent.* For example, $\{A, B, \{C, \{D, E\}, \{F, G\}\}\}$ is a nested sequent (where the letters stand for signed formulas). This is usually written as $A, B, [C, [D, E], [F, G]]$.

When presenting nested sequent rules, one makes use of a notion of *hole.* Think of $\{\ \}$ as a peculiar way of writing a special propositional letter. It is allowed to appear at most once in a nested sequent, but not as part of a more complex formula. For instance, $T\, A \wedge B, [F\, C, [T\, D \vee E, \{\ \}]]$ is a nested sequent with a hole. Schematically, it is abbreviated as $\Gamma\{\ \}$ and is referred to as a *context.* The result of 'filling' the hole with a specific signed formula, say Z, is written as $\Gamma\{Z\}$. In our example we get $T\,A \wedge B, [F\,C, [T\,D \vee E, Z]]$. More generally, a hole may be filled with multiple signed formulas, one or more nested sequents, or some combination of these. One can even allow filling with a nested sequent having a hole of its own, though we won't need such a complication.

We now give nested sequent rules for K. As with prefixed tableaus, we use signed formulas, and do not require negation normal form. Again, this is a minor point. In the following, $\Gamma\{\ \}$ is an arbitrary context; this will not be repeated each time.

Nested Sequent Axioms A nested sequent axiom is $\Gamma\{T\,X, F\,X\}$. If X is atomic, we say the axiom is *atomic*. It is common to make this a requirement.

Nested sequent proofs begin with axioms and end with a sequent containing only $T\,X$, where X is the formula being proved. In between, the following rules can be used.

Classical Nested Sequent Rules

$$\frac{\Gamma\{\alpha_1\} \quad \Gamma\{\alpha_2\}}{\Gamma\{\alpha\}} \qquad \frac{\Gamma\{\beta_1, \beta_2\}}{\Gamma\{\beta\}} \qquad \frac{\Gamma\{F\,X\}}{\Gamma\{T\,\neg X\}} \qquad \frac{\Gamma\{T\,X\}}{\Gamma\{F\,\neg X\}}$$

Modal Nested Sequent Rules

$$\frac{\Gamma\{[\nu_0]\}}{\Gamma\{\nu\}} \qquad \frac{\Gamma\{\pi, [\pi_0, \ldots]\}}{\Gamma\{\pi, [\ldots]\}}$$

In Example 2.3 we gave an instance of a prefixed tableau proof. Here is a corresponding nested sequent proof of the same formula. Think of the comma as disjunction and nested boxes as being necessitated. The connection between Example 2.3 and Example 7.2 is that one is the other turned over, with the roles of T and F reversed, and with nesting corresponding to prefix extension. Details can be found in [11].

Example 7.2.

$$\frac{\frac{\frac{\frac{\frac{\frac{F\,\Box(P \supset Q, F\,\Box P, [T\,P, F\,P, T\,Q] \quad F\,\Box(P \supset Q), F\,\Box P, [F\,Q, F\,P, T\,Q]}{F\,\Box(P \supset Q), F\,\Box P, [F\,P \supset Q, F\,P, T\,Q]}\,\alpha}{F\,\Box(P \supset Q), F\,\Box P, [F\,P, T\,Q]}\,\pi}{F\,\Box(P \supset Q), F\,\Box P, [T\,Q]}\,\pi}{F\,\Box(P \supset Q), F\,\Box P, T\,\Box Q}\,\nu}{F\,\Box(P \supset Q), T\,\Box P \supset \Box Q}\,\beta}{T\,\Box(P \supset Q) \supset (\Box P \supset \Box Q)}\,\beta$$

The nested sequent system for K, described above, can be turned into systems for other common modal logics with the addition of various rules. For instance, adding the following produces K4.

$$\frac{\Gamma\{\pi, [\pi, \ldots]\}}{\Gamma\{\pi, [\ldots]\}}$$

We follow a very different route here.

8 Indexed Nested Sequents

We now create an *indexed* version of nested sequents that will correspond to most set prefixed systems. There is a still more elaborate version that can handle all of them, and we discuss this briefly in Section 14. We begin with an explanation of why there are two versions.

Geach formulas semantically correspond to confluence properties. The iconic case is $\mathsf{G}^{1,1,1,1}$ which, semantically, is simply a diamond condition. The problems come with the trivial cases, since the Geach parameters are allowed to be 0. Speaking loosely, $\mathsf{G}^{k,l,m,n}$ says, about a frame, that a k length move and an m length move from the same starting point can be brought back together with an l length move and an n length move. But if both l and n are 0, we are not actually making any further moves. We are already at a single point, without doing anything more. If there were a fresh move to be made, we could place constraints on it. But if no move is to be made, we discover that constraints already exist, and so any notational representation we already have must be modified. It is this need to modify previously existing formal representations that forces us to more complex machinery. We avoid the complexity until after we have dealt with the simpler situations. Specifically, for the time being *we do not allow both l and n to be 0 in $\mathsf{G}^{k,l,m,n}$*, which covers most cases, and certainly all the standard ones. A more general approach will be discussed in Section 14.

We use *nested sequents* as sketched in Section 7, but instead of one kind of nested sequent we need a *family* of them. Each boxed sequent has an *index*, a non-negative integer, assigned to it using a notation convention illustrated by the following example: $\overset{3}{[}T\,A, F\,B, T\,C\overset{3}{]}$ is a boxed sequent with index 3. A top level sequent is not nested, but by convention we give it an index of 0 which does not turn up explicitly in the notation. We refer to the sequent calculi introduced here as *indexed nested sequents.*

Here is some informal motivation. With conventional nested sequents, and with the version here as well, a move from a sequent to a boxed sequent contained in it corresponds to a semantic move from a possible world to an accessible possible world. Loosely speaking, the overall sequent nesting structure represents edges in a directed graph or Kripke frame, with each nested sequent representing a possible world. But now the indexing machinery has the following function: *all sequents with the same index represent the same possible world.* Thus the tree structure inherent in nesting is supplemented to allow for multiple paths to the same world.

We now give the indexed sequent rules. Many of them have the same appearance as with ordinary nested sequents, except that sequent indexes are now present. We repeat rules here for convenience, even if they appear unchanged.

Indexed Nested Sequent Axioms An indexed nested sequent axiom is any sequent of the form $\Gamma\{T\,X, F\,X\}$. If X is atomic, the axiom is *atomic.*.

Classical Indexed Nested Sequent Rules

$$\frac{\Gamma\{\alpha_1\} \quad \Gamma\{\alpha_2\}}{\Gamma\{\alpha\}} \qquad \frac{\Gamma\{\beta_1, \beta_2\}}{\Gamma\{\beta\}} \qquad \frac{\Gamma\{F\,X\}}{\Gamma\{T\,\neg X\}} \qquad \frac{\Gamma\{T\,X\}}{\Gamma\{F\,\neg X\}}$$

Modal Indexed Nested Sequent Rules

$$\frac{\Gamma\{\overset{a}{[}\nu_0\overset{a}{]}\}}{\Gamma\{\nu\}} \qquad \frac{\Gamma\{\pi, \overset{a}{[}\pi_0, \ldots\overset{a}{]}\}}{\Gamma\{\pi, \overset{a}{[}\ldots\overset{a}{]}\}}$$

provided index a does not appear in the consequent

Next we have analogs of the tableau Set Prefix Union and Set Prefix Continuation Rules. These are most easily stated in words rather than symbolically. The first is a counterpart to the Set Prefix Union Rule.

Formula Contraction, FC If a signed formula occurs more than once in sequents with the same index, one occurrence can be deleted.

Here are two examples of the Formula Contraction Rule in use. The first reduces sequent size, eliminating an occurrence of $T\,C$, which appears in two boxed sequents with index 1. The second reduces a boxed sequent to empty—for this example, recall that the top level has index 0.

$$\frac{T\,A, F\,B, \overset{1}{[}T\,C, F\,D\overset{1}{]}, \overset{2}{[}T\,E, \overset{1}{[}T\,C, F\,G\overset{1}{]}\overset{2}{]}}{T\,A, F\,B, \overset{1}{[}T\,C, F\,D\overset{1}{]}, \overset{2}{[}T\,E, \overset{1}{[}F\,G\overset{1}{]}\overset{2}{]}}$$

$$\frac{T\,A, F\,B, \overset{1}{[}T\,C, F\,D, \overset{0}{[}T\,A\overset{0}{]}\overset{1}{]}}{T\,A, F\,B, \overset{1}{[}T\,C, F\,D, \overset{0}{[}\ \overset{0}{]}\overset{1}{]}}$$

Motivation for the Formula Contraction Rule is straightforward. In effect, no information is lost in deleting one signed formula occurrence, because the same information was recorded twice and so can be recovered in principle.

Sequent Contraction, SC If two boxed sequents with the same index, one empty $\overset{a}{[}\,\overset{a}{]}$ and one not necessarily empty $\overset{a}{[}\cdots\overset{a}{]}$, occur in sequents with the same index, $\overset{a}{[}\,\overset{a}{]}$ can be deleted.

The Sequent Contraction Rule (which is a counterpart of the Set Prefix Continuation Rule) is somewhat more complex. As noted earlier, a sequent can be thought of as designating a possible world. The formulas it contains are true at that possible world. Nesting structure of sequents corresponds to accessibility. Sameness of index for two sequents tells us they designate the same possible world. Of course information about formula truth is non-existent for an empty indexed sequent; nesting structure and index are all that matter for it. But in the premise of the Sequent Contraction Rule, information about nesting structure for an empty boxed sequent is postulated to be represented elsewhere, so information can be safely removed. Here is an example of the Sequent Contraction Rule. Two boxed sequents have index 1, and both occur in sequents with index 0 (recall, the top level sequent has index 0 by convention).

$$\frac{T\,A, \overset{1}{[}T\,B, F\,C\overset{1}{]}, \overset{0}{[}T\,D, \overset{1}{[}\ \overset{1}{]}\overset{0}{]}}{T\,A, \overset{1}{[}T\,B, F\,C\overset{1}{]}, \overset{0}{[}T\,D\overset{0}{]}}$$

The Geach Scheme we give is not a rule, but a rule generator. It creates rules for all of $\mathsf{G}^{k,l,m,n}$ where $k, l, m, n \geq 0$ *but not both* l *and* n *can be 0.* Direct notation for an arbitrary nested sequent can be quite cluttered. To avoid this problem we introduce some special notation in order to simplify presentation of the general form of the Geach Rule. In Section 9 we generate some instances of particular interest and present examples of the rules being used. When we come to specific cases, the special notation will no longer be needed.

Special Notation and Terminology Suppose $\Gamma\{\ \}$ is a context with a hole. We associate an index with this, which we refer to as the *index of the context.* If the hole is at the top level, the index of the context is 0. Otherwise the index is that of the boxed sequent in which the hole directly appears.

We use $\mathbb{N}$ for the set of non-negative integers, and $\mathbb{N}^n$ for the set of n-tuples over $\mathbb{N}$. Similarly, $\mathbb{S}$ is the set of indexed sequents (including the empty sequent) and $\mathbb{S}^n$ is the set of n-tuples of indexed sequents. We allow n to be 0, where the only member of $\mathbb{N}^0$ or $\mathbb{S}^0$ is $\langle\ \rangle$. We use ϵ to denote the empty sequent. It is understood that if S is a sequent, the sequent S, ϵ is the same as S.

For each $n \in \mathbb{N}$ we inductively define a mapping $\mathbb{SEQ}$ from $\mathbb{N}^n \times \mathbb{S}^n$ to sequents as follows.

$$\mathbb{SEQ}(\langle\ \rangle, \langle\ \rangle) = \epsilon$$

$$\mathbb{SEQ}(\langle a_1, a_2, \ldots, a_n\rangle, \langle S_1, S_2, \ldots, S_n\rangle) = \overset{a_1}{[}\, S_1, \mathbb{SEQ}(\langle a_2, \ldots, a_n\rangle, \langle S_2, \ldots, S_n\rangle)\,\overset{a_1}{]}$$

The following example should make the notation rather obvious. Suppose $a, b \in \mathbb{N}$ and $A, B \in \mathbb{S}$—non-negative integers, and sequents. Then:

$$\begin{aligned}
\mathbb{SEQ}(\langle a, b\rangle, \langle A, B\rangle) &= \overset{a}{[}A, \mathbb{SEQ}(\langle b\rangle, \langle B\rangle)\overset{a}{]} \\
&= \overset{a}{[}A, \overset{b}{[}B, \mathbb{SEQ}(\langle\ \rangle, \langle\ \rangle)\overset{ba}{]]} \\
&= \overset{a}{[}A, \overset{b}{[}B, \epsilon\overset{ba}{]]} \\
&= \overset{a}{[}A, \overset{b}{[}B\overset{ba}{]]}
\end{aligned}$$

Now we can give the general form of the Geach Scheme, in a reasonably readable way.

Indexed Nested Sequent Geach Scheme for $\mathbf{G}^{k,l,m,n}$ where not both l and n are 0.

$$\frac{\Gamma\{\mathbb{SEQ}(\langle a_1, ..., a_k, c_1, ..., c_l\rangle, \langle A_1, ..., A_k, \epsilon, ..., \epsilon\rangle), \mathbb{SEQ}(\langle b_1, ..., b_m, d_1, ..., d_n\rangle, \langle B_1, ..., B_m, \epsilon, ..., \epsilon\rangle)\}}{\Gamma\{\mathbb{SEQ}(\langle a_1, ..., a_k\rangle, \langle A_1, ..., A_k\rangle), \mathbb{SEQ}(\langle b_1, ..., b_m\rangle, \langle B_1, ..., B_m\rangle)\}}$$

The following conditions must be met.

1. (Distinctness Condition) All members of $c_1, \ldots, c_l$ must be distinct, and similarly for all members of $d_1, \ldots, d_n$.

2. (Confluence and Newness Conditions) if both of $\langle a_1, \ldots, a_k, c_1, \ldots, c_l\rangle$ and $\langle b_1, \ldots, b_m, d_1, \ldots, d_n\rangle$ are non-empty, they must have the same last terms, and apart from last terms, $c_1, \ldots, c_l$ and $d_1, \ldots, d_n$ must not overlap, nor may any c_i or d_i appear in the consequent (below the line).

3. (Degenerate Confluence and Newness Conditions) If one of $\langle a_1, ..., a_k, c_1, ..., c_l\rangle$ and $\langle b_1, ..., b_m, d_1, ..., d_n\rangle$ is empty, the last term of the non-empty one must be the index of the context. No c_i or d_i may appear in the consequent, except for the index of the context. (Not both sequences can be empty because not both l and n can be 0.)

9 Indexed Nested Sequent Examples

We give several specific examples of the Geach Scheme $\mathsf{G}^{k,l,m,n}$, and illustrate their use.

Example 9.1. ***Geach Rule for*** $\boldsymbol{G}^{0,1,0,0}$**,** $\Box X \supset X$ *Since only $l > 0$ we are in a very simple version of case 3 of the Geach Scheme. It reduces to the following.*

$$\frac{\Gamma\{\mathbb{SEQ}(\langle c_1\rangle, \langle\epsilon\rangle), \mathbb{SEQ}(\langle\ \rangle, \langle\ \rangle)\}}{\Gamma\{\mathbb{SEQ}(\langle\ \rangle, \langle\ \rangle), \mathbb{SEQ}(\langle\ \rangle, \langle\ \rangle)\}}$$

where c_1 is the index of the context. Expanding this, the rule is

$$\frac{\Gamma\{\overset{c_1}{[}\ \overset{c_1}{]}\}}{\Gamma\{\ \}}$$

where c_1 is the index of the context $\Gamma\{\ \}$.

Example 9.2. ***Geach Rule for*** $\boldsymbol{G}^{0,1,2,0}$**,** $\Box X \supset \Box\Box X$. *Now $l = 1$ and $m = 2$ with $k = n = 0$, so we are in case 2. The rule is the following.*

$$\frac{\Gamma\{\mathbb{SEQ}(\langle c_1\rangle, \langle\epsilon\rangle), \mathbb{SEQ}(\langle b_1, b_2\rangle, \langle B_1, B_2\rangle)\}}{\Gamma\{\mathbb{SEQ}(\langle\ \rangle, \langle\ \rangle), \mathbb{SEQ}(\langle b_1, b_2\rangle, \langle B_1, B_2\rangle)\}}$$

where we must have $c_1 = b_2$. Writing b_2 for both, this gives us the following.

$$\frac{\Gamma\{\overset{b_2}{[}\ \overset{b_2}{]}, \overset{b_1}{[} B_1, \overset{b_2}{[} B_2 \overset{b_2}{]}\overset{b_1}{]}\}}{\Gamma\{\overset{b_1}{[} B_1, \overset{b_2}{[} B_2 \overset{b_2}{]}\overset{b_1}{]}\}}$$

Proof of $\Box\Diamond\Diamond P \supset \Box\Diamond P$ using Geach $\mathbf{G}^{0,1,2,0}$ *This corresponds to set prefixed tableau Example 4.1.*

$$
\begin{array}{c}
F\,\Box\Diamond\Diamond P, \overset{1}{[}T\,\Diamond P, \overset{3}{[}T\,P, F\,P\overset{3}{]}, \overset{23}{[}[F\,P]\overset{321}{]}]\\
\hline
F\,\Box\Diamond\Diamond P, \overset{1}{[}T\,\Diamond P, \overset{3}{[}T\,P\overset{3}{]}, \overset{23}{[}[F\,P]\overset{321}{]}]\\
\hline
F\,\Box\Diamond\Diamond P, \overset{1}{[}T\,\Diamond P, \overset{3}{[}\,\overset{3}{]}, \overset{23}{[}[F\,P]\overset{321}{]}]\\
\hline
F\,\Box\Diamond\Diamond P, \overset{1}{[}T\,\Diamond P, \overset{23}{[}[F\,P]\overset{321}{]}]\\
\hline
F\,\Box\Diamond\Diamond P, \overset{1}{[}T\,\Diamond P, \overset{2}{[}F\,\Diamond P\overset{21}{]}]\\
\hline
F\,\Box\Diamond\Diamond P, \overset{1}{[}T\,\Diamond P, F\,\Diamond\Diamond P\overset{1}{]}\\
\hline
F\,\Box\Diamond\Diamond P, \overset{1}{[}T\,\Diamond P\overset{1}{]}\\
\hline
F\,\Box\Diamond\Diamond P, T\,\Box\Diamond P\\
\hline
T\,\Box\Diamond\Diamond P \supset \Box\Diamond P
\end{array}
\begin{array}{l}
\\ FC \\ \\ \pi \\ \\ G^{0,1,2,0} \\ \\ \nu \\ \\ \nu \\ \\ \pi \\ \nu \\ \beta
\end{array}
$$

Example 9.3. ***Geach Rule for $\mathbf{G}^{1,1,0,0}$, $\Diamond\Box X \supset X$*** *The rule scheme with $k = 1$, $l = 1$ and $m = n = 0$ is in case 3 and is the following, where c_1 is the index of the context.*

$$
\frac{\Gamma\{\mathbb{SEQ}(\langle a_1, c_1\rangle, \langle A_1, \epsilon\rangle), \mathbb{SEQ}(\langle\ \rangle, \langle\ \rangle)\}}{\Gamma\{\mathbb{SEQ}(\langle a_1\rangle, \langle A_1\rangle), \mathbb{SEQ}(\langle\ \rangle, \langle\ \rangle)\}}
$$

Writing this in conventional form, we have the following rule.

$$
\frac{\Gamma\{\overset{a_1}{[}A_1, \overset{c_1}{[}\ \overset{c_1}{]}\overset{a_1}{]}\}}{\Gamma\{\overset{a_1}{[}A_1\overset{a_1}{]}\}}
$$

where c_1 is the index of the context.

Proof of $\Diamond\Box P \supset \Box P$ using Geach $\mathsf{G}^{0,1,2,0}$ and $\mathsf{G}^{1,1,0,0}$ *This corresponds to Set Prefixed Example 4.3.*

$$\dfrac{\overset{1}{[}F\Box P, \overset{02}{[[}\ \overset{20}{]]}, \overset{2}{[}F P, T P\overset{21}{]]}, \overset{2}{[}T P\overset{2}{]}}{\overset{1}{[}F\Box P, \overset{02}{[[}\ \overset{20}{]]}, \overset{2}{[}F P\overset{21}{]]}, \overset{2}{[}T P\overset{2}{]}}\ \mathit{FC}$$

$$\dfrac{\overset{1}{[}F\Box P, \overset{02}{[[}\ \overset{20}{]]}, \overset{2}{[}F P\overset{21}{]]}, \overset{2}{[}T P\overset{2}{]}}{\overset{1}{[}F\Box P, \overset{02}{[[}\ \overset{20}{]]}, \overset{2}{[}\ \overset{21}{]]}, \overset{2}{[}T P\overset{2}{]}}\ \pi$$

$$\dfrac{\overset{1}{[}F\Box P, \overset{02}{[[}\ \overset{20}{]]}, \overset{2}{[}\ \overset{21}{]]}, \overset{2}{[}T P\overset{2}{]}}{\overset{1}{[}F\Box P, \overset{02}{[[}\ \overset{201}{]]]}, \overset{2}{[}T P\overset{2}{]}}\ G^{0,1,2,0}$$

$$\dfrac{\overset{1}{[}F\Box P, \overset{02}{[[}\ \overset{201}{]]]}, \overset{2}{[}T P\overset{2}{]}}{\overset{1}{[}F\Box P, \overset{0}{[}\ \overset{01}{]]}, \overset{2}{[}T P\overset{2}{]}}\ \mathit{FC}$$

$$\dfrac{\overset{1}{[}F\Box P, \overset{0}{[}\ \overset{01}{]]}, \overset{2}{[}T P\overset{2}{]}}{T\Box P, \overset{1}{[}F\Box P, \overset{0}{[}\ \overset{01}{]]}}\ \nu$$

$$\dfrac{T\Box P, \overset{1}{[}F\Box P, \overset{0}{[}\ \overset{01}{]]}}{T\Box P, \overset{1}{[}F\Box P\overset{1}{]}}\ G^{1,1,0,0}$$

$$\dfrac{\dfrac{T\Box P, \overset{1}{[}F\Box P\overset{1}{]}}{F\Diamond\Box P, T\Box P}\ \nu}{T\Diamond\Box P \supset \Box P}\ \beta$$

Example 9.4. ***Geach Rule for $G^{1,1,1,1}$, $\Diamond\Box X \supset \Box\Diamond X$*** *We are in case 2, and the rule scheme is the following, where $c_1 = d_1$, and may not occur in the consequent.*

$$\frac{\Gamma\{\mathbb{SEQ}(\langle a_1, c_1\rangle, \langle A_1, \epsilon\rangle), \mathbb{SEQ}(\langle b_1, d_1\rangle, \langle B_1, \epsilon\rangle)\}}{\Gamma\{\mathbb{SEQ}(\langle a_1\rangle, \langle A_1\rangle), \mathbb{SEQ}(\langle b_1\rangle, \langle B_1\rangle)\}}$$

Writing c_1 for both c_1 and d_1 we have the following.

$$\frac{\Gamma\{\overset{a_1}{[}A_1, \overset{c_1}{[}\ \overset{c_1 a_1}{]\,]}, \overset{b_1}{[}B_1, \overset{c_1}{[}\ \overset{c_1 b_1}{]\,]}\}}{\Gamma\{\overset{a_1}{[}A_1\overset{a_1}{]}, \overset{b_1}{[}B_1\overset{b_1}{]}\}}$$

where c_1 does not occur in the consequent. In Example 4.4 we gave a set prefixed tableau proof of $(\Diamond\Box P \wedge \Diamond\Box Q) \supset \Diamond(P \wedge Q)$ using Geach $G^{0,1,2,0}$ and $G^{1,1,1,1}$ rules. That corresponds to an indexed nested sequent proof, but it is rather long, so we omit it. Constructing it is a good exercise.

10 Design Decisions

Set prefixed tableaus and indexed nested sequents can be translated into each other, but in both directions some extra work needs to be done beyond a line for line translation. This is because we have made differing decisions in formulating the two systems. We have, in a sense, made tableau rules more maximal while allowing sequent rules to be more minimal, and this needs some comment.

A set prefixed tableau branch is closed if it contains $\Sigma\, T\, X$ and $\Sigma\, F\, X$ for some path set Σ and some formula X. As item 6 in the proof of Theorem 6.1 shows, instead of involving an entire path *set* Σ, it would have been sufficient to say we have closure when $\sigma\, T\, X$ and $\sigma\, F\, X$ both occur on the branch, for some path *sequence* σ. We could even have broadened things to allow closure if $\sigma_1\, T\, X$ and $\sigma_2\, F\, X$ both occur on the branch, where $\sigma_1 \sim \sigma_2$, using the equivalence relation defined in item 1 of the proof of Theorem 6.1. Since a path sequence corresponds to a nesting pattern, the first alternative is our indexed nested sequent axiom condition, which amounts to requiring that both $T\, X$ and $F\, X$ appear in the same nested sequent. The alternative involving the equivalence relation corresponds to requiring that both $T\, X$ and $F\, X$ occur in nested sequents having the same index, which is more generous than we have allowed. A sequent condition more directly corresponding to tableau branch closure would be a requirement that $T\, X$ and $F\, X$ both occur in *every* nested sequent having a given index, and this is more restrictive than we have required.

The comments we just made about our choices for tableau branch closure and sequent axioms have analogs for other rules as well. Design decisions were made. Why did we make the choices we did? Admittedly our choices were somewhat arbitrary, but some of our motivation was this. For tableaus, the most extreme version of rules would have involved bringing in the equivalence relation $\sim$. This can get complicated when constructing a tableau. Instead we thought of the set prefix itself as the place where information about $\sim$ could be explicitly stored, and wrote the tableau rules accordingly. On the other hand, when working with nested sequents the $\sim$ relation turns into, simply, having the same index, and this is an easy thing to recognize. Our actual choices were along these lines, but we did not take the extreme positions.

11 On Proving Completeness and Soundness

One can prove completeness for a proof system S_1 directly, or by showing some other system S_2, known to be complete, embeds into it. Similarly for soundness, though now the embedding is the other way around. If each of S_1 and S_2 embeds into the other in some direct way the two are, in some sense, notational variants of each other. In this sense Smullyan style tableaus and Gentzen sequents, for classical logic, are notational variants. Prefixed tableaus as in Section 2 and nested sequents as in Section 7 are notational variants—shown in [11]. Here we have chosen a mixed strategy. We show completeness of an indexed nested sequent system by showing that the corresponding set prefixed tableau system embeds into it. It is possible to show an embedding the other way around, establishing soundness, but it is complicated by the fact that more than one set prefixed signed formula corresponds to an indexed nested sequent, and so proofs cannot be translated line by line but must be handled globally. The same issue came up in [11], and was dealt with. Here, with an already long paper, we avoid the problem by showing soundness directly.

An anonymous referee for this paper pointed out a significant different direction that could be taken. Indexed nested sequents can be translated into a subsystem of a labelled sequent calculus of [23], G3K with appropriate rules added. We have decided not to make use of this here, for two reasons. The first is to keep the present paper self-contained. The second is that [23] is heavily proof-theoretic in nature, while we prefer a more semantic approach. Nonetheless, it is an important observation for which I thank the referee. I briefly sketch the idea in this section. I assume familiarity with [23], though in order to minimize inessentials I assume labeled sequents have been reformulated in a one-sided version, rather than the two-sided one that appears in that paper.

A labeled sequent consists of *relational atoms* of the form xRy where x and y are variables, and *labeled formulas* of the form $x{:}A$, where x is a variable and A is a modal formula. Intuitively, variables represent possible worlds. One can translate an indexed nested sequent to a labeled sequent as follows. First assign to each non-negative integer n a unique variable x_n—do this once and for all. Then, for a given indexed nested sequent, we build up a labeled sequent in the following way. If a boxed subsequent with index j occurs directly inside a subsequent with index i, add the relational atom $x_i R x_j$. And if signed formula Z occurs directly in a subsequent with index i, add $x_i{:}Z^*$, where $(T\,X)^* = X$ and $(F\,X)^* = \neg X$.

Consider Example 9.4, which gives the indexed nested sequent rule for $\mathsf{G}^{1,1,1,1}$. To simplify things we assume the context $\Gamma\{\ \}$ is empty, and we omit some subscripts. The upper and lower indexed nested sequents appearing in the rule translate as follows.

$$\overset{a}{[}A, \overset{c}{[}\ \overset{c}{]}\overset{a}{]}, \overset{b}{[}B, \overset{c}{[}\ \overset{c}{]}\overset{b}{]} \quad \text{becomes} \quad x_0Rx_a, x_aRx_c, x_0Rx_b, x_bRx_c, x_a{:}A^*, x_b{:}B^*$$

$$\overset{a}{[}A\overset{a}{]}, \overset{b}{[}B\overset{b}{]} \quad \text{becomes} \quad x_0Rx_a, x_0Rx_b, x_a{:}A^*, x_b{:}B^*$$

Now the indexed nested sequent rule instance

$$\dfrac{\overset{a}{[}A, \overset{c}{[}\,\overset{c}{]}\overset{a}{]}, \overset{b}{[}B, \overset{c}{[}\,\overset{c}{]}\overset{b}{]}}{\overset{a}{[}A\overset{a}{]}, \overset{b}{[}B\overset{b}{]}}$$
$$c \text{ not in consequent}$$

becomes

$$\dfrac{x_0Rx_a, x_aRx_c, x_0Rx_b, x_bRx_c, x_a{:}A^*, x_b{:}B^*}{x_0Rx_a, x_0Rx_b, x_a{:}A^*, x_b{:}B^*}$$
$$x_c \text{ not in consequent}$$

Except for the shift to one-sided sequents, this is the rule for Directedness from [23].

We do not follow this direction any further here, but the labeled sequents corresponding to indexed nested sequents clearly constitute a natural class, as do those corresponding to nested sequents without indexes.

12 Translating Tableaus to Sequents: Sequent Completeness

We prove completeness for indexed nested sequents by showing that set prefixed tableaus translate into them. In [11] we showed that ordinary prefixed tableau systems embedded into ordinary nested sequent systems. That work applies here as well, with a few additions and modifications to take care of indexes. Throughout this section we assume we have a set prefixed tableau system with some designated group Geach of Geach Rules, and a corresponding indexed nested sequent calculus with sequent counterparts of Geach. Exact details don't matter, except that in the tableau and sequent Geach Rules for $\mathsf{G}^{k,l,m,n}$ it is assumed that not both l and n are 0.

It is probably easiest to follow the tableau to sequent translation if we begin with an example. The example is small, and does not illustrate some of the complications that can arise. We discuss these complications afterwards. We have chosen Example 4.1, which contains a set prefixed tableau proof of $\Box\Diamond\Diamond P \supset \Box\Diamond P$ using $\mathsf{G}^{0,1,2,0}$. For convenience, we repeat the tableau here. In it 2 and 3 are from 1 by Classical α; 4 is from 3 by Modal π; 5 is from 2 by Modal ν; 6 is from 5 and 7 is from 6 by Modal π; 8 is from 4 and 7 by $\mathsf{G}^{0,1,2,0}$; 9 is from 4 by Modal ν; and 10 is from

7 and 8 by Set Prefix Union.

$$\begin{array}{lll}
\{\epsilon\} & F\,\Box\Diamond\Diamond P \supset \Box\Diamond P & 1.\\
\{\epsilon\} & T\,\Box\Diamond\Diamond P & 2.\\
\{\epsilon\} & F\,\Box\Diamond P & 3.\\
\{1\} & F\,\Diamond P & 4.\\
\{1\} & T\,\Diamond\Diamond P & 5.\\
\{1.2\} & T\,\Diamond P & 6.\\
\{1.2.3\} & T\,P & 7.\\
\{1.4, 1.2.3\} & T\,\top & 8.\\
\{1.4, 1.2.3\} & F\,P & 9.\\
\{1.4, 1.2.3\} & T\,P & 10.
\end{array} \quad (1)$$

Tableaus are backward reasoning systems while sequents are forward reasoning. The first step in our translation is to turn the tableau proof over and reverse the signs, turning it into a kind of forward reasoning proof. Many tableau rules are single-use, and once these are applied we can think of the set prefixed signed formula to which they are applied as being removed from the tableau branch. For instance, $\Sigma\,\alpha$ generates $\Sigma\,\alpha_1$ and $\Sigma\,\alpha_2$, and having been used, we can think of $\Sigma\,\alpha$ as no longer available. When turning a tableau over and reversing signs, α becomes β, with α_1 and α_2 becoming β_1 and β_2. Now we can think of $\Sigma\,\beta_1$ and $\Sigma\,\beta_2$ as generating $\Sigma\,\beta$, which was not present before (corresponding to single-use $\Sigma\,\alpha$ becoming unavailable for further use). Of course this does not apply to tableau rules that are not single-use. Here are a few of the inverted tableau rules as examples. We omit the rest, which are straightforward.

$$\frac{\Sigma\,\alpha_1 \quad \Sigma\,\alpha_2}{\Sigma\,\alpha} \qquad \frac{\Sigma\,\beta_1, \Sigma\,\beta_2}{\Sigma\,\beta} \qquad \frac{\Sigma.n\,\nu_0}{\Sigma\,\nu}\ n \text{ not in consequent} \qquad \frac{\Sigma\,\pi, \Sigma.n\,\pi_0}{\Sigma\,\pi}\ \Sigma.n \text{ in consequent}$$

After applying this inversion process to (1) we get (2), where we have numbered lines so they correspond to the numbering in the tableau. This is a proof in a forward reasoning system which we call *dual tableaus*. Formulas on tableau branches act conjunctively—all set prefixed formulas on a branch are understood as holding simultaneously. Lines in dual tableau (2) act disjunctively—at least one of the set prefixed formulas on a line holds. Tableaus close if a branch contains both $\Sigma\,T\,A$ and $\Sigma\,F\,A$—9 and 10 in (1). In (2) an axiom is a line containing both $\Sigma\,F\,A$ and $\Sigma\,T\,A$. And so on. A fuller discussion of the analogous step for ordinary prefixed tableaus can be found in Section 4 of [11].

$$\dfrac{\dfrac{\dfrac{\dfrac{\dfrac{\dfrac{\dfrac{\dfrac{\dfrac{\{\epsilon\}\,F\,\Box\Diamond\Diamond P, \{1.2.3\}\,F\,P, \{1\}\,T\,\Diamond P, \{1.4, 1.2.3\}\,F\,\top, \{1.4, 1.2.3\}\,T\,P, \{1.4, 1.2.3\}\,F\,P \quad \langle 10\rangle}{\{\epsilon\}\,F\,\Box\Diamond\Diamond P, \{1.2.3\}\,F\,P, \{1\}\,T\,\Diamond P, \{1.4, 1.2.3\}\,F\,\top, \{1.4, 1.2.3\}\,T\,P \quad \langle 9\rangle}}{\{\epsilon\}\,F\,\Box\Diamond\Diamond P, \{1.2.3\}\,F\,P, \{1\}\,T\,\Diamond P, \{1.4, 1.2.3\}\,F\,\top \quad \langle 8\rangle}}{\{\epsilon\}\,F\,\Box\Diamond\Diamond P, \{1.2.3\}\,F\,P, \{1\}\,T\,\Diamond P \quad \langle 7\rangle}}{\{\epsilon\}\,F\,\Box\Diamond\Diamond P, \{1.2\}\,F\,\Diamond P, \{1\}\,T\,\Diamond P \quad \langle 6\rangle}}{\{\epsilon\}\,F\,\Box\Diamond\Diamond P, \{1\}\,F\,\Diamond\Diamond P, \{1\}\,T\,\Diamond P \quad \langle 5\rangle}}{\{\epsilon\}\,F\,\Box\Diamond\Diamond P, \{1\}\,T\,\Diamond P \quad \langle 4\rangle}}{\{\epsilon\}\,F\,\Box\Diamond\Diamond P, \{\epsilon\}\,T\,\Box\Diamond P \quad \langle 2,3\rangle}}{\{\epsilon\}\,T\,\Box\Diamond\Diamond P \supset \Box\Diamond P \quad \langle 1\rangle} \tag{2}$$

The next step is to convert each line of a dual tableau from a collection of set prefixed signed formulas into an indexed nested sequent. This is basically simple, though a formal description is somewhat technical. The following is a variant of a similar translation from [11], except that now we must take sets and indexes into account.

Definition 12.1. *Let P be a set of prefixed signed formulas (note, not* set *prefixed, but prefixed by a path sequence in the traditional sense).*

1. *For each n let $P^n = \{\sigma\, Z \mid n.\sigma\, Z \in P\}$. (Recall that $n.\sigma$ is the prefix σ with n added at the beginning.)*

2. *Let $P^\circ = \{Z \mid \epsilon\, Z \in P\} \cup \left\{ \overset{n}{[}(P^n)^\circ\overset{n}{]} \mid P^n \neq \emptyset \right\}$. (In $\overset{n}{[}\cdots\overset{n}{]}$ here, the n is not actually a proper index—indexes will come later. Call it a* pseudo-index.*)*

For a set prefixed signed formula, $\Sigma\, Z$, the flattening *of it is $\{\sigma\, Z \mid \sigma \in \Sigma\}$. If S is a collection of set prefixed signed formulas, the* flattening *of S is the union of the flattenings of its members.*

Finally, for a collection S of set prefixed signed formulas, by S° we mean P°, where P is the flattening of S, but with all occurrences of $F\top$ deleted from P°.

For example, suppose $S = \{\{\epsilon\}\,F\,\Box\Diamond\Diamond P, \{1.2.3\}\,F\,P, \{1\}\,T\,\Diamond P, \{1.4, 1.2.3\}\,F\,\top\}$, item $\langle 8\rangle$ in (2). The flattening of S is

$$P = \{\epsilon\,F\,\Box\Diamond\Diamond P, 1.2.3\,F\,P, 1\,T\,\Diamond P, 1.4\,F\,\top, 1.2.3\,F\,\top\}.$$

Then:

$$
\begin{aligned}
P^\circ &= F\,\Box\Diamond\Diamond P, \overset{1}{[}\{2.3\,F\,P, \epsilon\,T\,\Diamond P, 4\,F\,\top, 2.3\,F\,\top\}^\circ\overset{1}{]} \\
&= F\,\Box\Diamond\Diamond P, \overset{1}{[}T\,\Diamond P, \overset{2}{[}\{3\,F\,P, 3\,F\,\top\}^\circ\overset{2}{]}, \overset{4}{[}\{\epsilon\,F\,\top\}^\circ\overset{4}{]}\overset{1}{]} \\
&= F\,\Box\Diamond\Diamond P, \overset{1}{[}T\,\Diamond P, \overset{2}{[}\overset{3}{[}\{\epsilon\,F\,P, \epsilon\,F\,\top\}^\circ\overset{3}{]}\overset{2}{]}, \overset{4}{[}F\,\top\overset{4}{]}\overset{1}{]} \\
&= F\,\Box\Diamond\Diamond P, \overset{1}{[}T\,\Diamond P, \overset{2}{[}\overset{3}{[}F\,P, F\,\top\overset{3}{]}\overset{2}{]}, \overset{4}{[}F\,\top\overset{4}{]}\overset{1}{]}
\end{aligned}
$$

Removing occurrences of $F\,\top$, we have the following.

$$
S^\circ = F\,\Box\Diamond\Diamond P, \overset{1}{[}T\,\Diamond P, \overset{2}{[}\overset{3}{[}F\,P\overset{3}{]}\overset{2}{]}, \overset{4}{[}\ \overset{4}{]}\overset{1}{]}
$$

The last step is to replace the pseudo-indexes by real indexes. We find it useful to do this as a two-step process. Note that pseudo-index 2 occurs in S° inside brackets having pseudo-index 1. We replace 2 with 1.2 to make this observation explicit, and we do something similar for each pseudo-index. We get the following.

$$
F\,\Box\Diamond\Diamond P, \overset{1}{[}T\,\Diamond P, \overset{1.2}{[}\ \overset{1.2.3}{[}\ F\,P\ \overset{1.2.3}{]}\ \overset{1.2}{]}, \overset{1.4}{[}\ \overset{1.4}{]}\overset{1}{]} \tag{3}
$$

For path sequences appearing in tableau (1), let us write $\sigma_1 \sim \sigma_2$ if σ_1 and σ_2 appear in the same path set, and let $\sim^*$ be the transitive closure of $\sim$. In this example $\sim^*$ is the same as $\sim$, but it may not happen generally. The relation $\sim$ is essentially the same as the one introduced in item 1 of the proof of Theorem 6.1. However, that relation was an equivalence relation because we were working with tableaus in which all possible rule applications were made. We are not making this assumption now since proofs may terminate before everything possible has been done, so transitive closure must be added explicitly.

Inspection of (1) shows the relation $\sim^*$ has 4 equivalences classes: $\{\epsilon\}$, $\{1\}$, $\{1.2\}$, and $\{1.4, 1.2.3\}$. We assign non-negative integers to these classes, making sure to assign 0 to the class containing ϵ, though otherwise things are arbitrary. Say we make the following assignment.

$$
\begin{aligned}
\{\epsilon\} &\rightarrow 0 \\
\{1\} &\rightarrow 1 \\
\{1.2\} &\rightarrow 2 \\
\{1.4, 1.2.3\} &\rightarrow 3
\end{aligned}
$$

We make this replacement in (3), getting the following nested sequent.

$$F\,\Box\Diamond\Diamond P, \overset{1}{[}T\,\Diamond P, \overset{23}{[[}F\,P\overset{32}{]]}, \overset{3}{[}\ \overset{31}{]]} \tag{4}$$

We began with the set $S = \{\{\epsilon\}\,F\,\Box\Diamond\Diamond P, \{1.2.3\}\,F\,P, \{1\}\,T\,\Diamond P, \{1.4, 1.2.3\}\,F\,\top\}$, and wound up with (4). We call (4) the *translate* of S, and write it as $\mathcal{T}(S)$. (The assignment of integers to equivalence classes is an implicit parameter of $\mathcal{T}$ and is assumed to be held constant throughout the translation of an entire proof.)

Now, apply translation $\mathcal{T}$ to every line of dual tableau (2). We get the following.

$$\begin{array}{cl}
F\,\Box\Diamond\Diamond P, \overset{123}{[[[}F\,P, T\,P\overset{32}{]]}, T\,\Diamond P, \overset{3}{[}T\,P, F\,P\overset{31}{]]} \quad \langle 10\rangle & \\
\hline
F\,\Box\Diamond\Diamond P, \overset{123}{[[[}F\,P, T\,P\overset{32}{]]}, T\,\Diamond P, \overset{3}{[}T\,P\overset{31}{]]} \quad \langle 9\rangle & \mathsf{FC} \\
\hline
F\,\Box\Diamond\Diamond P, \overset{123}{[[[}F\,P\overset{32}{]]}, T\,\Diamond P, \overset{3}{[}\ \overset{31}{]]} \quad \langle 8\rangle & \pi + \mathsf{FC} \\
\hline
F\,\Box\Diamond\Diamond P, \overset{123}{[[[}F\,P\overset{32}{]]}, T\,\Diamond P\overset{1}{]} \quad \langle 7\rangle & \mathsf{G}^{0.1.2.0} \\
\hline
F\,\Box\Diamond\Diamond P, \overset{12}{[[}F\,\Diamond P\overset{2}{]}, T\,\Diamond P\overset{1}{]} \quad \langle 6\rangle & \nu \\
\hline
F\,\Box\Diamond\Diamond P, \overset{1}{[}F\,\Diamond\Diamond P, T\,\Diamond P\overset{1}{]} \quad \langle 5\rangle & \nu \\
\hline
F\,\Box\Diamond\Diamond P, \overset{1}{[}T\,\Diamond P\overset{1}{]} \quad \langle 4\rangle & \pi \\
\hline
F\,\Box\Diamond\Diamond P, T\,\Box\Diamond P \quad \langle 2, 3\rangle & \nu \\
\hline
T\,\Box\Diamond\Diamond P \supset \Box\Diamond P \quad \langle 1\rangle & \beta
\end{array} \tag{5}$$

As it stands (5) is not quite an indexed nested sequent proof, but it is close. Line $\langle 10\rangle$ in (5) is overkill as an axiom, but it is required by the translation we have chosen. The passage from $\langle 10\rangle$ to $\langle 9\rangle$ is justified by Formula Contraction, the counterpart of the tableau Prefix Union rule which served to justify 10 in (1). The problem comes with the passage from $\langle 9\rangle$ to $\langle 8\rangle$, which is not justified as it stands since the π rule for sequents only eliminates π_0 from a single sequent, and not from a family of sequents having the same index. But we can achieve this effect by first using Formula Contraction and then the π rule, as shown below.

$$\begin{array}{cl}
F\,\Box\Diamond\Diamond P, \overset{123}{[[[}F\,P, T\,P\overset{32}{]]}, T\,\Diamond P, \overset{3}{[}T\,P\overset{31}{]]} \quad \langle 9\rangle & \\
\hline
F\,\Box\Diamond\Diamond P, \overset{123}{[[[}F\,P\overset{32}{]]}, T\,\Diamond P, \overset{3}{[}T\,P\overset{31}{]]} & \mathsf{FC} \\
\hline
F\,\Box\Diamond\Diamond P, \overset{123}{[[[}F\,P\overset{32}{]]}, T\,\Diamond P, \overset{3}{[}\ \overset{31}{]]} \quad \langle 8\rangle & \pi
\end{array}$$

Now that we have seen an example, a general discussion is already partly motivated. Each Set Prefixed Tableau Rule application converts to an Indexed Nested Sequent Rule application, or sometimes to a series of them. Most rules fall into a similar, simple pattern, and we discuss these rules first. Consider the Set Prefixed α Rule: replace $\Sigma\,\alpha$ with $\Sigma\,\alpha_1$ and $\Sigma\,\alpha_2$ on a tableau branch. We invert, and switch T and F signs. The sign switch turns α signed formulas into β signed formulas. Next, for the dual tableau stage, we have a rule: replace $\Sigma\,\beta_1, \Sigma\,\beta_2$ with $\Sigma\,\beta$. When we turn this into an indexed nested sequent step, using the translation of Definition 12.1, each path sequence in Σ gives rise to a nested sequent but each nested sequent has the same index since all come from members of the same set Σ. So the schematic form of the result is the following.

$$\frac{\overset{i}{[}\cdots,\beta_1,\beta_2\overset{i}{]},\ldots,\overset{i}{[}\cdots,\beta_1,\beta_2\overset{i}{]}}{\overset{i}{[}\cdots,\beta\overset{i}{]},\ldots,\overset{i}{[}\cdots,\beta\overset{i}{]}}$$

This is not, directly, an allowed step in an indexed nested sequent proof, but it can obviously be replaced with a series of steps, each involving a single β rule application, one for each of the nested sequents displayed above the line.

Our discussion of the Set Prefixed α Tableau Rule applies just as well to all the Classical Set Prefixed Rules, and to the Set Prefixed ν Rule. The π rule is different, however. It replaces $\Sigma\,\pi$ on a tableau branch with $\Sigma.n\,\pi_0$, where n is new. When creating the dual tableau this becomes the following: $\Sigma.n\,\nu_0$ can be replaced with $\Sigma\,\nu$, provided n does not occur in the conclusion. When converting to nested sequents, Σ becomes a collection of nested sequents with the same index, and $\Sigma.n$ becomes a collection of nested sequents appearing directly inside those, again with the same index. We thus have the following general form.

$$\frac{\overset{i}{[}\cdots,\overset{j}{[}\nu_0\overset{j}{]}\overset{i}{]},\ldots,\overset{i}{[}\cdots,\overset{j}{[}\nu_0\overset{j}{]}\overset{i}{]}}{\overset{i}{[}\cdots,\nu\overset{i}{]},\ldots,\overset{i}{[}\cdots,\nu\overset{i}{]}}$$

Unlike the cases discussed above, this cannot be replaced with multiple applications of the Set Prefixed ν Rule because of the requirement that index j not occur in the consequent. To get around this we make use of a familiar structural result. The proof is a straightforward induction on sequent proof length, and is omitted.

Theorem 12.2 (Weakening). *If Γ is a provable indexed nested sequent, and Γ' is like Γ but with signed formula Z added to some subsequent, then Γ' is also provable.*

With weakening available, we can proceed as follows. Using Theorem 12.2, if there is a proof of the indexed sequent

$$\overset{i}{[}\cdots,\overset{j}{[}\nu_0\overset{j}{]}\overset{i}{]},\ldots,\overset{i}{[}\cdots,\overset{j}{[}\nu_0\overset{j}{]}\overset{i}{]}$$

then there is also a proof of the following.

$$\overset{i}{[}\cdots,\nu,\overset{j}{[}\nu_0\overset{j}{]}\overset{i}{]},\ldots,\overset{i}{[}\cdots,\nu,\overset{j}{[}\nu_0\overset{j}{]}\overset{i}{]}$$

From this, using iterated applications of Formula Contraction, we derive

$$\overset{i}{[}\cdots,\nu,\overset{j}{[}\nu_0\overset{j}{]}\overset{i}{]},\ldots,\overset{i}{[}\cdots,\nu,\overset{j}{[}\,\overset{j}{]}\overset{i}{]}$$

where only a single occurrence of ν_0 in a j indexed boxed sequent remains. Next using iterated applications of Sequent Contraction, we derive

$$\overset{i}{[}\cdots,\nu,\overset{j}{[}\nu_0\overset{j}{]}\overset{i}{]},\ldots,\overset{i}{[}\cdots,\nu\overset{i}{]}$$

where $\overset{j}{[}\nu_0\overset{j}{]}$ in the leftmost boxed sequent is the only nested sequent indexed with j. Finally, using the Indexed Nested Sequent ν Rule we get the desired sequent.

$$\overset{i}{[}\cdots,\nu\overset{i}{]},\ldots,\overset{i}{[}\cdots,\nu\overset{i}{]}$$

Rather like the α rule discussed above, the Set Prefix Union Rule converts to iterated applications of Indexed Nested Sequent Formula Contraction, and Set Prefix Continuation to Indexed Nested Sequent Formula Contraction, followed by Indexed Nested Sequent Contraction.

The last item to discuss is Set Prefixed Geach Rules, which bring certain complications of their own. In order to keep notation down we again work with a specific rule example, $\mathsf{G}^{1,1,1,1}$, but it should be sufficiently representative. For convenience, we restate the tableau rule here.

$$\frac{\begin{array}{c}\sigma.a \in \Sigma_1 \text{ on branch}\\ \sigma.b \in \Sigma_2 \text{ on branch}\end{array}}{\begin{array}{c}(\Sigma_1.c \cup \Sigma_2.d)\, T\, \top\\ \text{where}\\ c \text{ and } d \text{ are new and distinct}\end{array}}$$

We translate this rule into an indexed nested sequent step, which will then need some amending. Using Definition 12.1, each path sequence is turned into a nested sequent with indexes assigned using equivalence classes, as outlined above. All nested sequents arising from path sequents in $\Sigma_1.c \cup \Sigma_2.d$ receive the same index, and similarly for path sequents in Σ_1 and Σ_2 themselves. Let us say nested sequents coming from path sequences in Σ_1 have index p, those coming from Σ_2 have index q, and those coming from $\Sigma_1.c \cup \Sigma_2.d$ have index r. It is important to note that, since c and d are new and distinct, no path sequence in $\Sigma_1.c \cup \Sigma_2.d$ can already appear on the tableau branch at this point, and so under translation *only* nested sequents corresponding to members of $\Sigma_1.c \cup \Sigma_2.d$ will have index r at this point, and all of these nested sequents will be empty since during translation $T\top$ becomes $F\top$ which is omitted.

We can think of σ as corresponding to the context for this rule application. Since the rule is inverted when translating from tableaus to sequents, this step becomes the *removal* of sequent items arising from $(\Sigma_1.c \cup \Sigma_2.d)\,T\top$. Since $\sigma.a \in \Sigma_1$ and $\sigma.b \in \Sigma_2$, before the inverted sequent rule is applied we must have nested sequents corresponding to path sequents $\sigma.a.c$ and $\sigma.b.d$ present, both empty, and with index r. That is, we must have $\overset{p}{[}\cdots,\overset{r}{[}\,\overset{r}{]}\overset{p}{]}$ and $\overset{q}{[}\cdots,\overset{r}{[}\,\overset{r}{]}\overset{q}{]}$. But Σ_1 and Σ_2 may have members besides $\sigma.a$ and $\sigma.b$, so the sequent translate of $(\Sigma_1.c\cup\Sigma_2.d)\,T\top$ may involve multiple occurrences of $\overset{r}{[}\,\overset{r}{]}$ besides the ones just displayed. Some of these sequents come from $\Sigma_1.c T\top$, and look like $\overset{p}{[}\cdots,\overset{r}{[}\,\overset{r}{]}\overset{p}{]}$ where index p derives from Σ_1 and the index r derives from $\Sigma_1.c$ which is part of $\Sigma_1.c \cup \Sigma_2.d$. The other sequents come from $\Sigma_2.d T\top$ and look like $\overset{q}{[}\cdots,\overset{r}{[}\,\overset{r}{]}\overset{q}{]}$. Thus the tableau rule application translates into the following sequent step.

$$\dfrac{\begin{array}{c}\Gamma\{\overset{p}{[}\cdots,\overset{r}{[}\,\overset{r}{]}\overset{p}{]},\overset{q}{[}\cdots,\overset{r}{[}\,\overset{r}{]}\overset{q}{]}\}\\ \text{various other occurrences of } \overset{p}{[}\cdots,\overset{r}{[}\,\overset{r}{]}\overset{p}{]}\\ \text{various other occurrences of } \overset{q}{[}\cdots,\overset{r}{[}\,\overset{r}{]}\overset{q}{]}\end{array}}{\begin{array}{c}\Gamma\{\overset{p}{[}\cdots\overset{p}{]},\overset{q}{[}\cdots\overset{q}{]}\}\\ \text{various other occurrences of } \overset{p}{[}\cdots\overset{p}{]}\\ \text{various other occurrences of } \overset{q}{[}\cdots\overset{q}{]}\end{array}}$$

This is not quite an application of the Indexed Sequent $\mathsf{G}^{1,1,1,1}$ rule, because of the possible presence of the other occurrences of $\overset{r}{[}\,\overset{r}{]}$, but it can be turned into a correct series of steps. First using the **Sequent Contraction Rule**, one by one, eliminate the other occurrences of $\overset{r}{[}\ \overset{r}{]}$ occurring in p indexed boxed sequents, but keeping the occurrence in the p indexed sequent in context $\Gamma\{\ \}$. Next, do the same with occurrences of $\overset{r}{[}\ \overset{r}{]}$ occurring in q indexed sequents. This leaves us with the following.

$$\begin{array}{c}\Gamma\{\overset{p}{[}\cdots,\overset{r}{[}\ \overset{r}{]}\overset{p}{]},\overset{q}{[}\cdots,\overset{r}{[}\ \overset{r}{]}\overset{q}{]}\}\\ \text{various other occurrences of } \overset{p}{[}\cdots\overset{p}{]}\\ \text{various other occurrences of } \overset{q}{[}\cdots\overset{q}{]}\end{array}$$

Finally, apply the sequent rule for $\mathsf{G}1,1,1,1$, and the desired conclusion is attained.

This completes discussion of the translation from tableaus to sequents, and establishes sequent completeness.

13 Sequent Soundness

We have shown that tableaus translate into sequents, and hence we have sequent completeness. Sequent soundness could be proved by providing a translation in the other direction, and then relying on tableau soundness. As we noted earlier, this is somewhat complex and we do not do it. In [4] the soundness proof for nested (but not indexed) sequents makes use of a translation from sequents into formulas called *corresponding formulas*. We do not follow this route either because the appropriate version of corresponding formula for indexed sequents involves nominals, and hybrid logics come into the picture, [3]. This is too much of a detour for present purposes. We have chosen to transfer our soundness proof from tableaus to sequents, which is an easier thing to describe than translating individual proofs. The analogy of the present soundness proof to that of Section 5 should be relatively clear.

For the following, recall we have extended the notion of truth at possible worlds to allow signed formulas; specifically, $\mathcal{M},\Gamma \Vdash T\,X$ if $\mathcal{M},\Gamma \Vdash X$, and $\mathcal{M},\Gamma \Vdash F\,X$ if $\mathcal{M},\Gamma \not\Vdash X$.

Definition 13.1 (Structural Mappings)**.** *Let $\mathcal{S}$ be an indexed nested sequent. We write $\mathcal{S}^*$ for the collection consisting of the top level sequent $\mathcal{S}$ itself, and all boxed subsequents. For each $\Gamma \in \mathcal{S}^*$, we write $i(\Gamma)$ for the index of Γ, and $\mathcal{I}(\mathcal{S})$ for $\{i(\Gamma) \mid \Gamma \in \mathcal{S}^*\}$.*

Let $\mathcal{M} = \langle \mathcal{G}, \mathcal{R}, \mathcal{V}\rangle$ be a Kripke model, as in Section 5. A mapping $s : \mathcal{I}(\mathcal{S}) \to \mathcal{G}$ is a structural mapping of $\mathcal{S}$ into $\mathcal{M}$ *provided that, for all $\Gamma, \Delta \in \mathcal{S}^*$, if $\Delta \in \Gamma$ then*

$s(i(\Gamma))\mathcal{R}s(i(\Delta))$.

Suppose s is a structural mapping of $\mathcal{S}$ into $\mathcal{M}$. For $\Gamma \in \mathcal{S}^*$, we say signed formula $Z \in \Gamma$ is true under s *if $\mathcal{M}, s(i(\Gamma)) \Vdash Z$. We say Γ itself is true under s if some signed formula belonging to Γ is true under s, or else some boxed sequent Δ belonging to Γ is true under s. We say Γ is* false under s *if it is not true under s.*

The definition of truth for an indexed nested sequent is recursive. Speaking loosely, an indexed nested sequent $\mathcal{S}$ will be true under a structural mapping s if it has a true signed formula member, or it has a true boxed sequent member, where this boxed sequent member will be true if it has a true signed formula member, or it has a true boxed sequent member, and so on. Then an alternate characterization is, $\mathcal{S}$ is true under s if some subsequent contains a true signed formula. Likewise $\mathcal{S}$ is false under s if every subsequent contains only false signed formulas.

Soundness will be an easy consequence of the following three Lemmas.

Lemma 13.2. *Assume s is a structural mapping of $\mathcal{S}$ into $\mathcal{M}$.*

α **Case:** *Suppose $\mathcal{S} = \Gamma\{\alpha\}$. Then $\Gamma\{\alpha\}$ is true under s if and only if both $\Gamma\{\alpha_1\}$ and $\Gamma\{\alpha_2\}$ are true under s.*

β **Case:** *Suppose $\mathcal{S} = \Gamma\{\beta\}$. Then $\Gamma\{\beta\}$ is true under s if and only if $\Gamma\{\beta_1, \beta_2\}$ is true under s.*

$T\neg$ **Case:** *Suppose $\mathcal{S} = \Gamma\{T\neg X\}$. Then $\Gamma\{T\neg X\}$ is true under s if and only if $\Gamma\{F X\}$ is true under s.*

$F\neg$ **Case:** *Suppose $\mathcal{S} = \Gamma\{F\neg X\}$. Then $\Gamma\{F\neg X\}$ is true under s if and only if $\Gamma\{T X\}$ is true under s.*

π **Case:** *Suppose $\mathcal{S} = \Gamma\{\pi, \overset{a}{[}\ldots\overset{a}{]}\}$. Then $\Gamma\{\pi, \overset{a}{[}\ldots\overset{a}{]}\}$ is true under s if and only if $\Gamma\{\pi, \overset{a}{[}\pi_0, \ldots\overset{a}{]}\}$ is true under s.*

FC Case: *Suppose $\mathcal{S}$ follows from $\mathcal{S}'$ using the* Formula Contraction Rule*. Then $\mathcal{S}$ is true under s if and only if $\mathcal{S}'$ is true under s.*

SC Case: *Suppose $\mathcal{S}$ follows from $\mathcal{S}'$ using the* Sequent Contraction Rule*. Then $\mathcal{S}$ is true under s if and only if $\mathcal{S}'$ is true under s.*

Proof. We discuss the β case in some detail, the SC case briefly, and omit discussion of the rest. The remaining cases are similar and can be left to the reader.

First, note that if s is a structural mapping for $\Gamma\{\beta\}$, it is also a structural mapping for $\Gamma\{\beta_1, \beta_2\}$, since both of these indexed nested sequents have the same indexes and the same pattern of nesting. Similar observations apply to the other cases as well, though the SC case requires a bit more of an argument. For SC, the premise and consequent of a Sequent Contraction Rule application have the same indexes, but the nesting pattern changes a bit because the premise contains both $\overset{a}{[}\,\overset{a}{]}$ and $\overset{a}{[}\cdots\overset{a}{]}$ while the consequent contains $\overset{a}{[}\cdots\overset{a}{]}$ but not $\overset{a}{[}\,\overset{a}{]}$. But SC requires that both $\overset{a}{[}\cdots\overset{a}{]}$ and $\overset{a}{[}\,\overset{a}{]}$ appear in sequents having the same index, and this is enough to ensure that any structural mapping for the consequent of an SC rule application is also a structural mapping for the premise.

Suppose $\Gamma\{\beta_1, \beta_2\}$ is true under s. We show $\Gamma\{\beta\}$ is also true under s. By our assumption, some subsequent of $\Gamma\{\beta_1, \beta_2\}$ contains a signed formula that is true under s. If this signed formula is not one of β_1 or β_2, the same signed formula will also be present in $\Gamma\{\beta\}$, and so $\Gamma\{\beta\}$ will be true under s. If one of β_1 or β_2 is true under s then β is also true, and so again $\Gamma\{\beta\}$ will be true under s. The converse direction is similar. □

Lemma 13.3 (ν Case). *Suppose $\mathcal{S}_1$ and $\mathcal{S}_2$ are indexed nested sequents and $\mathcal{S}_2$ follows from $\mathcal{S}_1$ using the ν Rule. If there is a structural mapping of $\mathcal{S}_2$ into $\mathcal{M}$ under which $\mathcal{S}_2$ is false, it can be extended to a structural mapping of $\mathcal{S}_1$ into $\mathcal{M}$ under which $\mathcal{S}_1$ is false.*

Proof. Let $\mathcal{S}_1 = \Gamma\{\overset{a}{[}\nu_0\overset{a}{]}\}$ and $\mathcal{S}_2 = \Gamma\{\nu\}$, where a does not occur in $\mathcal{S}_2$, so that $\mathcal{S}_2$ follows from $\mathcal{S}_1$ using the ν Rule. Assume s_2 is a structural mapping of $\Gamma\{\nu\}$ into $\mathcal{M}$ under which $\Gamma\{\nu\}$ is false. We define a mapping s_1 as follows. We know that a does not occur in $\Gamma\{\nu\}$ but apart from a, $\Gamma\{\nu\}$ has the same indexes as $\Gamma\{\overset{a}{[}\nu_0\overset{a}{]}\}$, and with the same nesting pattern. On these indexes, let s_1 agree with s_2. Say the index of the context in $\Gamma\{\ \}$ is b. Since $\Gamma\{\nu\}$ is false under s_2 every subsequent contains only false signed formulas, so in particular, $\mathcal{M}, s_2(b) \not\Vdash \nu$. Then there is some possible world w of $\mathcal{M}$ so that $s_2(b)\mathcal{R}w$ and $\mathcal{M}, w \not\Vdash \nu_0$. Set $s_1(a) = w$. It is straightforward to check that s_1 is a structural mapping of $\Gamma\{\overset{a}{[}\nu_0\overset{a}{]}\}$ into $\mathcal{M}$, and $\Gamma\{\overset{a}{[}\nu_0\overset{a}{]}\}$ is false under s_1. □

Lemma 13.4 ($\mathsf{G}^{k,l,m,n}$ Case). *Suppose $\mathcal{S}_1$ and $\mathcal{S}_2$ are indexed nested sequents and $\mathcal{S}_2$ follows from $\mathcal{S}_1$ using Geach rule $\mathsf{G}^{k,l,m,n}$, where not both l and n are 0. Let $\mathcal{M}$ be a Kripke model that meets the semantic condition for $\mathsf{G}^{k,l,m,n}$. If there is a structural mapping s_2 of $\mathcal{S}_2$ into $\mathcal{M}$, it can be extended to a structural mapping s_1 of $\mathcal{S}_1$ into $\mathcal{M}$ so that $\mathcal{S}_1$ is true under s_1 if and only if $\mathcal{S}_2$ is true under s_2.*

Proof. Let us say that $\Gamma\{\ \}$ is a context with index e,

$$\begin{aligned}\mathcal{S}_1 = \Gamma\{&\mathbb{SEQ}(\langle a_1, \ldots, a_k, c_1, \ldots, c_l\rangle, \langle A_1, \ldots, A_k, \epsilon, \ldots, \epsilon\rangle),\\ &\mathbb{SEQ}(\langle b_1, \ldots, b_m, d_1, \ldots, d_n\rangle, \langle B_1, \ldots, B_m, \epsilon, \ldots, \epsilon\rangle)\}\\ \mathcal{S}_2 = \Gamma\{&\mathbb{SEQ}(\langle a_1, \ldots, a_k\rangle, \langle A_1, \ldots, A_k\rangle), \mathbb{SEQ}(\langle b_1, \ldots, b_m\rangle, \langle B_1, \ldots, B_m\rangle)\}\end{aligned}$$

and s_2 is a structural mapping of $\mathcal{S}_2$ into $\mathcal{M}$, where $\mathcal{M}$ meets Geach condition $\mathsf{G}^{k,l,m,n}$ (with at least one of l or n non-zero). We construct a structural mapping s_1.

By the **Geach Rule** conditions, no members of either $\{c_1, \ldots, c_l\}$ or $\{d_1, \ldots, d_n\}$ may appear in $\mathcal{S}_2$. All other indexes are common to both $\mathcal{S}_1$ and $\mathcal{S}_2$ and with the same nesting pattern involved. On these other indexes let s_1 agree with s_2.

At least one of $\{c_1, \ldots, c_l\}$ and $\{d_1, \ldots, d_n\}$ is non-empty. To keep the language simple in the following discussion, we will talk as if both sets were non-empty. Obvious modifications cover the cases where one is empty.

Recall that the index of the context in $\Gamma\{\ \}$ is e. By the definition of $\mathbb{SEQ}$ from Section 8, in $\mathcal{S}_2$ there is a nested tower of boxed sequents with indices $a_1, \ldots, a_k$, all contained in a sequent whose index is e. Likewise there is another nested tower of boxed sequents with indices $b_1, \ldots, b_m$, also contained in the same sequent with index is e. Since s_2 is a structural mapping, in $\mathcal{M}$ there is a path of possible worlds, $s_2(e)$, $s_2(a_1)$, …, $s_2(a_k)$, with each accessible from its predecessor (this is a path with k edges). There is a similar path with m edges, whose nodes are $s_2(e)$, $s_2(b_1)$, …, $s_2(b_m)$. (Note that by our s_1 definition so far, s_1 and s_2 agree on each a_i and b_i, and on e.) Since $\mathcal{M}$ meets the semantic condition for $\mathsf{G}^{k,l,m,n}$, there are paths of length l and n, starting at $s_2(a_k)$ and $s_2(b_m)$ respectively and ending at a common possible world, let us call it ω. By conditions on the **Geach Rule**, members of $\{c_1, \ldots, c_l\}$ must be distinct, and similarly for members of $\{d_1, \ldots, d_n\}$; c_l and d_n must be the same, and otherwise $\{c_1, \ldots, c_l\}$ and $\{d_1, \ldots, d_n\}$ must not overlap. This is enough to allow us to define s_1 on $c_1, \ldots, c_l$ to be the possible worlds (after the first) on the path from $s_2(a_k) = s_1(a_k)$ to ω, and likewise to define s_1 on $d_1, \ldots, d_n$ to be the possible worlds on the path from $s_2(b_m) = s_1(b_m)$ to ω. We have now defined s_1 on all indexes in $\mathcal{S}_1$.

We leave it to the reader to verify that s_1 is a structural mapping from $\mathcal{S}_1$ into $\mathcal{M}$. Now, in $\mathbb{SEQ}(\langle a_1, \ldots, a_k, c_1, \ldots, c_l\rangle, \langle A_1, \ldots, A_k, \epsilon, \ldots, \epsilon\rangle)$, the boxed subsequents with indexes c_1, …, c_l contain no signed formulas. Likewise in $\mathbb{SEQ}(\langle b_1, \ldots, b_m, d_1, \ldots, d_n\rangle, \langle B_1, \ldots, B_m, \epsilon, \ldots, \epsilon\rangle)$, boxed subsequents with indexes d_1, …, d_n contain no signed formulas. It follows that $\mathcal{S}_1$ and $\mathcal{S}_2$ contain the same signed formulas, and these occur in subsequents having the same indexes in both $\mathcal{S}_1$ and $\mathcal{S}_2$, and s_1 and s_2 agree on these indexes. Then if either of $\mathcal{S}_1$ or $\mathcal{S}_2$ had a subsequent with a true signed formula, both would have such, and so $\mathcal{S}_1$ and $\mathcal{S}_2$ must evaluate to the same truth value under s_1 and s_2 respectively. □

Theorem 13.5 (Soundness). *Assume we have an indexed nested sequent system with a set* Geach *of rules* $\mathsf{G}^{k,l,m,n}$ *with not both* l *and* n *being 0. Let* $\mathcal{K}$ *be the class of Kripke models whose frames satisfy the semantic conditions for* Geach*. If formula* X *has a sequent proof,* X *is true at every possible world of every model in* $\mathcal{K}$*.*

Proof. Suppose X has a sequent proof but in model $\mathcal{M}$ of $\mathcal{K}$, X is false at possible world w. We derive a contradiction.

Since X has $\mathcal{M}$ as a counter-model, we have $\mathcal{M}, w \not\Vdash X$ and hence $\mathcal{M}, w \not\Vdash T\,X$. A sequent proof of X is a tree of indexed nested sequents with axioms at the leaves and just $T\,X$ at the root—the last line of the proof. This root is an indexed nested sequent whose index is 0. Let s be a mapping with domain $\{0\}$ such that $s(0) = w$. This is a structural mapping of the root sequent of the proof into $\mathcal{M}$, and under it the root sequent is false.

Using Lemmas 13.2, 13.3, and 13.4 we see that if there is a structural mapping from some line of a nested sequent proof into Kripke model $\mathcal{M}$ under which that line is false, the same will be the case for at least one of the premises of that proof line. Since this is the case for the root, we can trace this upward and conclude that for at least one of the leaves of the proof tree, it is falsified by some structural mapping into $\mathcal{M}$. But leaves are axioms, and it is easy to see these are not falsifiable. □

14 The Missing Rules

We did not give nested sequent systems corresponding to all Geach formulas, $\Diamond^k\Box^l X \supset \Box^m\Diamond^n X$. We imposed a requirement that not both l and n could be 0. At the start of Section 8 we briefly discussed why such a restriction was needed. Essentially it comes down to this. In the nested sequent cases we allowed, no nested sequent already introduced in a proof ever needed its index modified. For the missing cases, modification is needed. We briefly sketch an idea for dealing with this, but we do not fully develop it here. (It may be that additional structural rules are needed to ensure completeness.) What is presented are suggestions, not conclusions. We hope others will find the proposal of enough interest to carry the investigation further.

For a more general indexed nested sequent system than those discussed earlier, allow sequents to have *multiple* indexes, or more properly, *sets as indexes.* For example,

$$\overset{\{0,2,3\}}{[}\quad A, B, C \quad\overset{\{0,2,3\}}{]}$$

is what we will call a *set indexed sequent*. We say $\{0, 2, 3\}$ is *the* index of this, while 2 is *an* index, as are 0 and 3. Think of all members of the set as names for the same possible world. In Section 11 we noted a connection with indexed nested sequents and labeled sequents. In fact, in the labelled sequent approach the machinery that must be added to treat the cases where both l and n are 0 is *equality*. Our use of sets as indexes clearly amounts to a version of equality. When using $\{0, 2, 3\}$, the possible worlds named by 0, 2, and 3 are equal.

As usual, a proof of X is a proof of the sequent consisting of only $T\,X$. We still assume that 0 is an index of the top level sequent. Axioms and propositional rules have the same form as before. Modal rules are almost the same. They are as follows.

Modal Set Indexed Nested Sequent Rules

$$\frac{\Gamma\{\overset{\{a\}}{[}\;\nu_0\;\overset{\{a\}}{]}\}}{\Gamma\{\nu\}}$$

provided a does not appear in the consequent

$$\frac{\Gamma\{\pi, \overset{s}{[}\pi_0, \ldots\overset{s}{]}\}}{\Gamma\{\pi, \overset{s}{[}\ldots\overset{s}{]}\}}$$

where s is any set index

Formula Contraction and Sequent Contraction have the same form as before, but "same index" should be interpreted to mean that set indexes share a member. And finally, using notation introduced in Section 8, we have the following, where the a_i and b_i are set indexes.

Set Indexed Nested Sequent Geach Scheme for $\mathsf{G}^{k,0,m,0}$, $\Diamond^k X \supset \Box^m X$

$$\frac{\Gamma\{\mathbb{SEQ}(\langle a_1, \ldots, a_{k-1}, a_k \cup b_m\rangle, \langle A_1, \ldots, A_k\rangle), \mathbb{SEQ}(\langle b_1, \ldots, b_{m-1}, a_k \cup b_m\rangle, \langle B_1, \ldots, B_m\rangle)\}}{\Gamma\{\mathbb{SEQ}(\langle a_1, \ldots, a_k\rangle, \langle A_1, \ldots, A_k\rangle), \mathbb{SEQ}(\langle b_1, \ldots, b_m\rangle, \langle B_1, \ldots, B_m\rangle)\}}$$

Here is an example using $\mathsf{G}^{2,0,0,0}$ (with axiom scheme $\Diamond\Diamond X \supset X$) and $\mathsf{G}^{3,0,0,0}$ (with axiom scheme $\Diamond\Diamond\Diamond X \supset X$). For these, the Geach Rules are as follows.

$$\frac{\Gamma\left\{\overset{a_1}{[}\cdots, \overset{a_2\cup b}{[}\cdots\overset{a_2\cup b}{]}\overset{a_1}{]}\right\}}{\Gamma\left\{\overset{a_1}{[}\cdots, \overset{a_2}{[}\cdots\overset{a_2}{]}\overset{a_1}{]}\right\}}$$

where b is the index of the context

$$\frac{\Gamma\left\{\overset{a_1}{[}\cdots, \overset{a_2}{[}\cdots, \overset{a_3\cup b}{[}\cdots\overset{a_3\cup b}{]}\overset{a_2}{]}\overset{a_1}{]}\right\}}{\Gamma\left\{\overset{a_1}{[}\cdots, \overset{a_2}{[}\cdots, \overset{a_3}{[}\cdots\overset{a_3}{]}\overset{a_2}{]}\overset{a_1}{]}\right\}}$$

where b is the index of the context

Using these rules, here is a sequent proof of $(\Diamond\Diamond A \wedge \Diamond X) \supset X$.

$$\begin{array}{c}
[^{\{1\}}\ [^{\{0,2\}}\ FA,\ [^{\{0,3\}}\ FX, TX\ ]^{\{0,3\}}\ ]^{\{0,2\}}\ ]^{\{1\}}, F\Diamond X, TX \\
\hline
[^{\{1\}}\ [^{\{0,2\}}\ FA,\ [^{\{0,3\}}\ FX\ ]^{\{0,3\}}\ ]^{\{0,2\}}\ ]^{\{1\}}, F\Diamond X, TX \\
\hline
[^{\{1\}}\ [^{\{0,2\}}\ FA,\ [^{\{3\}}\ FX\ ]^{\{3\}}\ ]^{\{0,2\}}\ ]^{\{1\}}, F\Diamond X, TX \\
\hline
[^{\{1\}}\ [^{\{0,2\}}\ FA, F\Diamond X\ ]^{\{0,2\}}\ ]^{\{1\}}, F\Diamond X, TX \\
\hline
[^{\{1\}}\ [^{\{0,2\}}\ FA\ ]^{\{0,2\}}\ ]^{\{1\}}, F\Diamond X, TX \\
\hline
[^{\{1\}}\ [^{\{2\}}\ FA\ ]^{\{2\}}\ ]^{\{1\}}, F\Diamond X, TX \\
\hline
[^{\{1\}}\ F\Diamond A\ ]^{\{1\}}, F\Diamond X, TX \\
\hline
F\Diamond\Diamond A, F\Diamond X, TX \\
\hline
F\Diamond\Diamond A \wedge \Diamond X, TX \\
\hline
T(\Diamond\Diamond A \wedge \Diamond X) \supset X
\end{array}
\quad
\begin{array}{l}
\mathsf{FC} \\ \mathsf{G}^{3,0,0,0} \\ \nu \\ \mathsf{FC} \\ \mathsf{G}^{2,0,0,0} \\ \nu \\ \nu \\ \beta \\ \beta
\end{array}$$

Since modal logics involving $\mathsf{G}^{n,0,0,0}$ are relatively rare, we conclude with a sketch of an axiomatic proof of $(\Diamond\Diamond A \wedge \Diamond X) \supset X$ using axiom schemes $\mathsf{G}^{2,0,0,0}$ and $\mathsf{G}^{3,0,0,0}$.

$$\begin{array}{rll}
\neg X & \supset \Box\Box\Box\neg X & \text{dual of } \mathsf{G}^{3,0,0,0} \\
(\Diamond\Diamond A \wedge \neg X) & \supset (\Diamond\Diamond A \wedge \Box\Box\Box\neg X) & \\
& \supset \Diamond\Diamond(A \wedge \Box\neg X) & \text{using } (\Diamond U \wedge \Box V) \supset \Diamond(U \wedge V) \\
& \supset (A \wedge \Box\neg X) & \text{using } \mathsf{G}^{2,0,0,0} \\
& \supset \Box\neg X &
\end{array}$$

We thus have $(\Diamond\Diamond A \wedge \neg X) \supset \Box\neg X$, from which $(\Diamond\Diamond A \wedge \Diamond X) \supset X$ follows easily.

15 Conclusion

We noted at the beginning of this paper that other approaches already provide tableau systems for the Geach logics. Then what might be gained from the approach introduced here? The key, we hope, lies in the fact that the tableau systems presented here do not expand the language. From this we have at least the possibility of two interesting consequences.

In [13] it is shown how ordinary prefixed tableaus and nested sequents can be used to prove interpolation constructively, in a uniform way, for all the propositional modal logics in the S5 modal cube. There is reason to hope that this can be pushed further using set prefixed tableaus, or indexed nested sequents, or the corresponding fragment of labeled sequent systems. Unlike [3], set prefixed tableaus and indexed nested systems do not expand the language of the logic itself, though they do expand the machinery used in formal proofs. It is possible that an investigation along present lines might give some proof-theoretic insights into which logics admit interpolation, and why.

Second, in unpublished work that builds on [12] we have shown that all logics in the Geach family have justification logic counterparts, with Realization Theorems connecting justification and modal versions. (See [1] for more on justification logics.) The proof is non-constructive. A number of modal logics have constructively proven Realization Theorems. One constructive approach, in [15], makes use of standard prefixed tableaus (and their equivalent, nested sequents) for the logics in the S5 modal cube. It is at least possible that set prefixed tableaus could allow a constructive proof of realization for the Geach family.

References

[1] Sergei Artemov and Melvin Fitting. Justification logic. In Edward N. Zalta, editor, *The Stanford Encyclopedia of Philosophy.* Fall 2012 edition, 2012.

[2] Patrick Blackburn, Maarten de Rijke, and Yde Venema. *Modal Logic.* Tracts in Theoretical Computer Science. Cambridge University Press, Cambridge, UK, 2001.

[3] Patrick Blackburn and Baldur ten Cate. Beyond pure axioms: Node creating rules in hybrid tableaux. In M. Marx, C. Areces, P. Blackburn, and U. Sattler, editors, *Proceedings of the 4th Workshop on Hybrid Logics (HyLo 2002)*, pages 21–35, 2002.

[4] Kai Brünnler. Deep sequent systems for modal logic. *Archive for Mathematical Logic*, 48(6):551–577, 2009.

[5] Kau Brünnler and Luts Straßburger. Modular sequent systems for modal logic. In Martin Giese and Arild Waaler, editors, *Automated Reasoning with Analytic Tableaux and Related Methods*, volume 5607 of *Lecture Notes in Artificial Intelligence*, pages 152–166. Springer, 2009.

[6] Brian F. Chellas. *Modal Logic, an introduction.* Cambridge University Press, 1980.

[7] Marcello D'Agostino, Dov Gabbay, Reiner Hähnle, and Joachim Posegga, editors. *Handbook of Tableau Methods.* Kluwer, Dordrecht, 1999.

[8] Melvin C. Fitting. Tableau methods of proof for modal logics. *Notre Dame Journal of Formal Logic*, 13:237–247, 1972.

[9] Melvin C. Fitting. *Proof Methods for Modal and Intuitionistic Logics.* D. Reidel Publishing Co., Dordrecht, 1983.

[10] Melvin C. Fitting. Modal proof theory. In Patrick Blackburn, Johan van Benthem, and Frank Wolter, editors, *Handbook of Modal Logic*, chapter 2, pages 85–138. Elsevier, 2007.

[11] Melvin C. Fitting. Prefixed tableaus and nested sequents. *Annals of Pure and Applied Logic*, 163:291–313, 2012. Available on-line at http://dx.doi.org/10.1016/j.apal.2011.09.004.

[12] Melvin C. Fitting. Justification logics and realization. Technical Report TR-2014004, CUNY Ph.D. Program in Computer Science, March 2014. `http://www.cs.gc.cuny.edu/tr/`.

[13] Melvin C. Fitting and Roman Kuznets. Modal interpolation via nested sequents. *Annals of Pure and Applied Logic*, 166:274–305, 2015.

[14] Dov M. Gabbay. *Labelled Deductive Systems*, volume I of *Oxford Logic Guides, 33*. Clarendon Press, 1996.

[15] Remo Goetschi and Roman Kuznets. Realization for justification logics via nested sequents: Modularity through embedding. *Annals of Pure and Applied Logic*, 163(9):1271–1298, September 2012.

[16] Rajeev Goré. *Tableau methods for modal and temporal logics*, pages 297–396, in [7].

[17] Rajeev Gore, Linda Postniece, and Alwen F Tiu. On the correspondence between display postulates and deep inference in nested sequent calculi for tense logics. *Logical Methods in Computer Science*, 7:1–38, 2011.

[18] Rajeev Goré and Revantha Ramanayake. Labelled tree sequents, tree hypersequents and nested (deep) sequents. In Thomas Bolander, Torben Braüner, Silvio Ghilardi, and Lawrence Moss, editors, *Advances in Modal Logic*, volume 9, pages 279–299. College Publications, 2012.

[19] Marcus Kracht. Power and weakness of the modal display calculus. In Heinrich Wansing, editor, *Proof Theory of Modal Logics*, pages 92–121. Kluwer, 1996.

[20] Sonia Marin and Lutz Straßburger. Label-free modular systems for classical and intuitionistic modal logics. In Rajeev Goré, Barteld Kooi, and Agi Kurucz, editors, *Advances in Modal Logic*, volume 10, pages 387–406. Advances in Modal Logic, College Publications, 2014.

[21] Fabio Massacci. Strongly analytic tableaux for normal modal logics. In Alan Bundy, editor, *Proceedings of CADE 12*, volume 814 of *Lecture Notes in Artificial Intelligence*, pages 723–737, Berlin, 1994. Springer-Verlag.

[22] Claudia Nalon, João Marcos, and Clare Dixon. Clausal resolution for modal logics of confluence. In *IJCAR 2014*, pages 322–336, 2014.

[23] Sara Negri. Proof analysis in modal logic. *Journal of Philosophical Logic*, 34:507–544, 2005.

Received May 2015

Retalis Language for Information Engineering in Autonomous Robot Software

Pouyan Ziafati[a,b]*, Mehdi Dastani[b], John-Jules Meyer[b],
Leendert van der Torre[a,c] and Holger Voos[a]
[a] *Centre for Security, Reliability and Trust (SnT), University of Luxembourg*
[b] *Intelligent Systems Group, Utrecht University*
[c] *Computer Science and Communications Research Unit, University of Luxembourg*
{Pouyan.Ziafati, Leon.Vandertorre, Holger.Voos}@uni.lu
{M.M.Dastani, J.J.C.Meyer}@uu.nl

Abstract

Robotic information engineering is the processing and management of data to create knowledge of the robot's environment. It is an essential robotic technique to apply AI methods such as situation awareness, task-level planning and knowledge-intensive task execution. Consequently, information engineering has been identified as a major challenge to make robotic systems more responsive to real-world situations. The *Retalis* language integrates *ELE* and *SLR*, two logic-based languages. *Retalis* is used to develop information engineering components of autonomous robots. In such a component, *ELE* is used for temporal and logical reasoning, and data transformation in flows of data. *SLR* is used to implement a knowledge base maintaining a history of events. *SLR* supports state-based representation of knowledge built upon discrete sensory data, management of sensory data in active memories and synchronization of queries over asynchronous sensory data. In this paper, we introduce eight requirements for robotic information engineering, and we show how *Retalis* unifies and advances the state-of-the-art research on robotic information engineering. Moreover, we evaluate the efficiency of *Retalis* by implementing an application for a NAO robot. *Retalis* receives events about the positions of objects with respect to the top camera of NAO robot, the transformation among the coordinate frames of

We would like to thank three anonymous reviewers for their valuable comments and suggestions to improve the quality of this paper. We would like to thank also Yehia El Rakaiby, Sergio Sousa, and Marc van Zee for their contributions in implementation or preparing the previous presentations of this work.

*Sponsored by *Fonds National de la Recherche Luxembourg (FNR)*.

NAO robot, and the location of the NAO robot in the environment. About one thousand and nine hundreds events per second are processed in real-time to calculate the positions of objects in the environment.

1 Introduction

Robotic information engineering is the processing and management of data to create knowledge of the robot's environment. In artificial intelligence (AI), knowledge of the environment is typically represented in symbolic form. Symbolic representation of knowledge is essential for robots with AI capabilities such as situation awareness, task-level planning, knowledge-intensive task execution and human interaction [66, 14, 81, 72, 50]. Challenges of robotic information engineering include the processing and management of incremental, discrete and asynchronous sensory data such as recognized faces[1] [25], objects[2] [9], gestures [69] and behaviors [59]. Data processing includes applying logical, temporal, spatial and probabilistic reasoning techniques [71, 41, 16, 50, 46, 30, 65].

Both *on-demand* and *on-flow* processing of sensory data are necessary for a timely extraction and dissemination of information in robot software. On-demand processing is the modeling and management of data in different memory profiles such as short, episodic and semantic memories [79, 80, 65, 70]. Memory profiles are accessed and processed when required. For example, a plan execution component requests the location of a previously observed object in order to find it. On-flow processing is the processing of sensory data on the fly in order to extract information about the environment. An example is the monitoring of smoke and temperature sensor readings in order to detect fire. A fire alarm should be generated if there is smoke and the temperature is high, observed by sensors in close proximity within a given time interval. A notification about fire detection is sent, for instance, to a plan execution component to react on it. We refer to receivers of the notifications as consumers.

On-demand processing includes the following requirements.

1. Memorizing: data should be recorded selectively to avoid overloading memory.
2. Forgetting: outdated data should be pruned to bound the amount of recorded data in memory.
3. Active memory: memory should notify consumers when it is updated with relevant information. In this way, consumers can access the information at their time of convenience.

[1]http://wiki.ROS.org/face_recognition
[2]http://wiki.ROS.org/object_recognition

4. State-based representation: knowledge about the state of the robot's environment should be derived from discrete observations.

On-flow processing includes the following requirements.

1. Even-driven and incremental processing: on-flow processing requires a real-time event-driven processing model. Relevant information should be derived as soon as it can be inferred from the sensory data received so far. Therefore, sensory data should be processed and reasoned about as soon as they are received by the system. Moreover, real-time processing of sensory data requires incremental processing techniques.
2. Temporal pattern detection and transformation: on-flow processing requires detecting temporal patterns in flow of data and transforming data into suitable representations. The detection and transformation of data patterns are required to correlate and aggregate sensory data and detect high-level events occurring in the robot's environment.
3. Subscription: information derived from on-flow processing of data should be disseminated selectively. This is needed, for instance, not to overload a plan execution component with irrelevant events.
4. Garbage collection: records of data should be kept as far as they can contribute to derive relevant information and pruned afterwards. In the fire alarm example, a detection of smoke needs to be kept for a specified time period. If a relevant sensor detects a high temperature during this period, a fire alarm is generated. The record of the detected smoke is disregarded afterwards.

The aim of this paper is to support robotic information engineering. A key concern to develop affordable, maintainable and reliable robot software is the support of re-usability in development [19, 20, 35]. Support of information engineering includes identifying the requirements and providing re-usable solutions to requirements. Re-usability advances robot software by reducing development, maintenance and benchmarking costs [35, 54, 55, 42, 13, 37]. A robotic language should support information engineering as follows. First, it should support implementation at a suitable level of abstraction. This includes a qualitative representation of temporal relations among events as opposed to specifying such relations by occurrence times of events. Second, it should support efficient implementation, because an incremental processing and management of sensory data requires specialized algorithms and implementation care. Third, it should have clear semantics to support the correctness of implementation. In particular, the language semantics should take the asynchronicity of data into account. Fourth, it should support AI reasoning techniques as its built-in functionalities or by integration of relevant tools and libraries. For instance, logical and spatial reasoning capabilities are often necessary to reason about the domain and common-sense knowledge, and spatial relations among objects.

Current systems do not support both on-flow and on-demand processing. The following examples illustrate the need to combine on-flow and on-demand processing. First, active memories generate events when the contents of their memories change. It is desirable that a consumer is able to subscribe for notification when a pattern of such changes occurs [80]. This requires an on-flow processing mechanism to process the memory events to detect relevant patterns of memory updates. Second, on-flow processing is needed for transforming data to a compact and suitable representation before recording it in memory. Third, simpler and more efficient implementation of some on-flow processing tasks can be achieved by mixing on-flow pattern recognition with on-demand querying of data in memory. In addition, on-flow and on-demand processing support of existing systems is limited. Open-source robotic software such as *ROS* [61] only facilitate flows of data among components. A state-of-the-art system is *DyKnow* [42, 38], which integrates multiple tools such as *C-SPARQL* [12] to support on-flow processing [27, 44, 39]. *C-SPARQL* does not support the expression of qualitative temporal relations among data or the filtering of data patterns based on their durations. Such capabilities are desirable, if not necessary, to capture complex data patterns [4]. In addition, on-flow processing in *DyKnow* requires semantic annotation of flows of data. Such semantic annotation is not provided in *ROS* software, widely used by the community, inducing programming overhead. Moreover, *DyKnow* is not available as open-source. Current on-demand processing systems often support either logical reasoning or active memory, but not both. An exception is the logic-based knowledge management system *ORO* [50, 49]. *ORO* supports active memory, but its support for the following on-demand requirements are limited. First, *ORO* does not support selectively memorizing data. All input data is recorded. Second, forgetting is limited to fixed memory profiles. It is not possible to specify forgetting policies based on types of data. Third, due to the open world assumption, representing and reasoning about dynamics of the robot's environment is difficult in *ORO*.

This paper introduces *Retalis* (*ETALIS*[3] [6, 5, 3] for Robotics) to supports on-flow and on-demand logical and temporal reasoning over sensory data and the state of robot's environment. *Retalis* is open source[4] and has been integrated in *ROS*. *Retalis* integrates the *Etalis* language for events (*ELE*)[5] for on-flow and develops the Synchronized Logical Reasoning language (*SLR*) [83] for on-demand processing.

[3]Event TrAnsaction Logic Inference System, http://code.google.com/p/etalis/

[4]https://github.com/procrob/Retalis

[5]*ETALIS* provides also the Event Processing SPARQL language (EP-SPARQL) [4] for event processing in Semantic Web applications.

By a seamless integration of these languages, *Retalis* supports the implementation of both on-flow and on-demand functionalities in one program.

The remainder of this paper is organized as follows. Section 2 presents a running example. Section 3 gives an overview of *Retalis*. Section 4 discusses on-flow processing requirements and describes *ELE*. Section 5 discusses on-demand processing requirements and presents *SLR*. Section 6 provides an evaluation of *Retalis*. Finally, Section 7 presents future work and concludes the paper.

2 Running Example

This section presents an example to illustrate the concepts and functionalities of *Retalis*. A robot is situated in a dynamic environment informing a person about the objects around it. The environment is described by a set of entities $e_1, e_2, \ldots$. These include the moving *base* of the robot, the pan-tilt 3D camera *cam* of the robot mounted on the *base*, a set of tables $table_1, table_2, \ldots$, a set of objects $o_1, o_2, \ldots$, a set of people $f_1, f_2, \ldots$, a set of attributes and a reference coordination frame *rcf*.

Figure 1 presents the robot's software components and their interactions. This figure should be read as follows. Directed arrows visualize asynchronous flows of data and two-way arrows represents request-response service calls. Asynchronous communications among the components are in the form of events. An event is a time-stamped piece of data formally defined in Section 4.1.

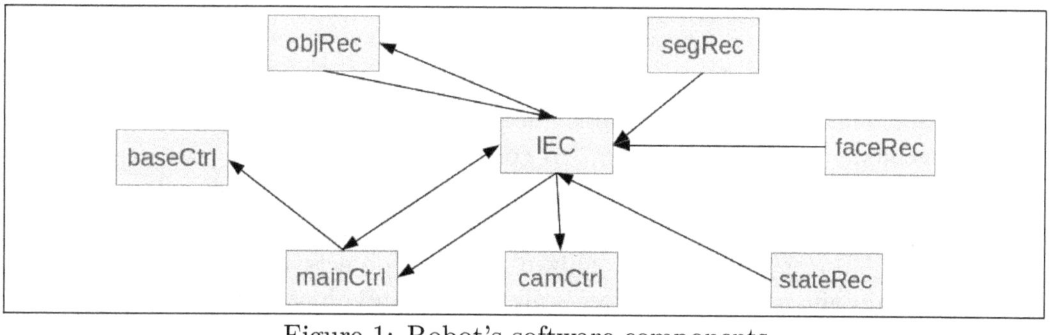

Figure 1: Robot's software components

The robot software includes the following components.

faceRec component: processes images from the camera, outputting $face(f_i, p_j)^t$ events. A $face(f_i, p_j)^t$ event represents the recognition of the face of f_i with confidence value p_j in a picture taken at time t.

segRec component: uses a real-time algorithm to process images from the camera into 3D point cloud data segments corresponding to individual objects. Such an algorithm is presented by Uckermann et al. [74]. The *segRec* component outputs $seg(o_i, c_j, p_k, l_g, pcl_h)^t$ events. Such an event represents the recognition of object o_i, with color c_j, with probability p_k, with relative position l_g to the *cam*, with the 3D point cloud data segment pcl_h recognized from a picture taken at time t. For events of the recognition of the same object segment over time, a unique identifier o_i is assigned using an anchoring and data association algorithm. Such an algorithm is presented by Elfring et al. [30].

objRec component: processes 3D point cloud data segments, outputting events of the form $obj(o_i, ot_j, p_k)^t$. Such an event represents the recognition of object type ot_j with probability p_k for object o_i recognized from a picture taken at time t.

stateRec component: localizes the robot. It outputs two types of events. A $tf(rcf, \text{base}, l_k)^t$ event represents the relative position between the reference coordination frame and the robot base at time t. A $tf(base, cam, l_k)^t$ event represents the relative position between the robot base and its camera at time t.

camCtrl and *baseCtrl* components: receive events of type $pos\_goal(l)$, each containing a position l to point the camera toward l or move the robot base to l, respectively.

IEC component: processes and manages events from *faceRec*, *segRec*, *stateRec* components. It detects reliable recognition of faces and objects and their movements to inform the *mainCtrl* component. Moreover, it positions objects in the reference coordination frame. In addition, it sends point cloud data of some objects to the *objRec* components to have their types recognized. The *IEC* component receives recognized types of objects from *objRec* as events and maintains the history of recognized faces and objects. It also controls the camera's position to follow a specific entity by sending perceived positions of the entity to *camCtrl*.

mainCtrl component: is responsible for interacting with the user. It moves the robot base by sending commands to the *baseCtrl* component. It receives events from *IEC* about the movements of objects to inform the user. The *mainCtrl* component queries *IEC* to answer the questions of the user.

3 Architectural Overview of *Retalis*

Retalis is a language for implementing Information Engineering Components (*IECs*) of autonomous robot systems. *IECs* are software components implementing a variety of information processing and management functionalities. *IEC*s are distributed independent components operating with other software components in parallel. *Retalis* does not impose any restriction on how components are structured in robot software.

Retalis represents and manipulates data as events. Events are time-stamped discrete pieces of data whose syntax is the same as *Prolog* ground terms [23, 53]. Events contain perceptual information such as a robot's position at a time or recognized objects in a picture. The meaning of events is domain-specific. The time-stamp of an event is a time point or a time interval referring to the occurrence time of the event. Events are time-stamped by the components generating them. [6] For example, the event *face('Neda',70)*^{28} could mean a recognition of Neda's face with 70% confidence in a picture taken at time 28 and the event *observed('Neda')*$^{\langle 28,49\rangle}$ could mean a frequent recognition of Neda's face in pictures taken during time interval [28,49]. An event containing information from processing of sensory data is usually time-stamped with the time at which the sensory data is acquired. This is usually different from the time point when the processing of the data is finished. A composite event generated from an occurrence of a pattern of other events is time-stamped based on the occurrence times of its composing events.

Retalis comprises two logic-based languages. The *ELE* language [6, 5, 3] supports on-flow processing and the *SLR* language [83] supports on-demand processing of data. In the *Retalis* program of an *IEC*, *ELE* generates composite events by detecting event patterns of interest in the input flow of events to the *IEC*. *SLR* is used to implement a knowledge base maintaining the history of some events. The knowledge base contains domain knowledge, including rules to reason about the recorded history. The flow of events processed by the *IEC* includes its input events and the composite events it generates. This means that composite events can in turn be used to detect other events. The robot software presented in Figure 1 includes one *IEC* component. Robot software can include a number of *IEC* components in order to modularize different information engineering tasks and to use distributed and parallel computing resources.

Figure 2 depicts the architecture of an *IEC*, including its logical components implemented in *Retalis*. This figure must be read as follows. Directed arrows visualize

[6] We assume all components share a central clock which is usually the clock of the computer running the components. If there is a network of computers running the components, time should be synchronized among them.

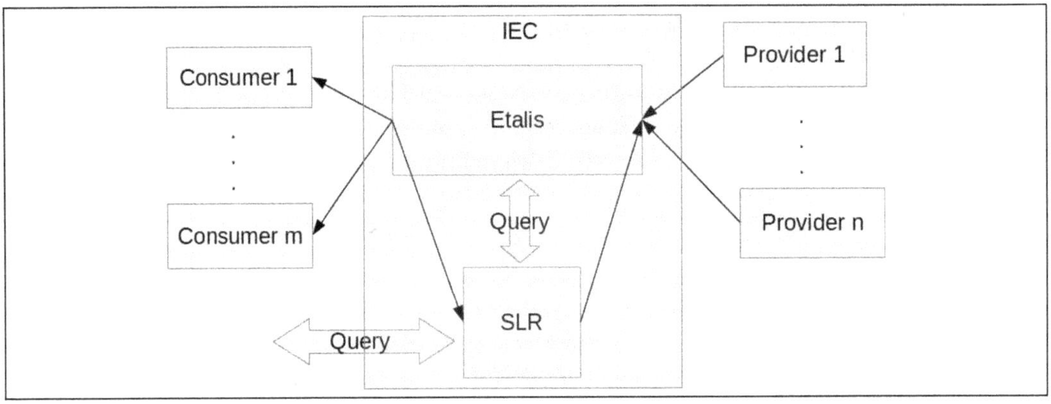

Figure 2: *IEC* architecture

asynchronous flows of events. Two-way arrows represent queries to *SLR* by *ELE* and external components.

Retalis supports the implementation of both synchronous and asynchronous interfaces among *IEC*s and other components. Asynchronous interaction is realized as follows. The *IEC* subscribes to events provided by *provider 1, .., provider n.* Moreover, *consumer 1, .., consumer n* subscribe to the *IEC* for types of events. The *Retalis* execution is event-driven. Input events are processed as they are received by the *IEC* to derive new events. When an event is processed, the event and resulting composite events are sent to interested consumers. The history of the input and derived events is also recorded in memory according to the *SLR* specification. *Retalis* specifications can be reconfigured at runtime. This includes the composite events to be detected, the producers the *IEC* is subscribed to, the subscriptions of consumers to the *IEC*, and the history of events maintained in memory.

Synchronous interactions between the *IEC* and other components are as follows. Components can query the domain knowledge and history of events in the *SLR* knowledge base. *Retalis* provides a request-response service to query *SLR*. *SLR* is a *Prolog*-based language, presented in Section 5.1 . The evaluation of a *SLR* query determines whether the query can be inferred from the knowledge base. The query evaluation may result in a variable substitution. The *IEC* can also access the functionalities of other software libraries or components. Function calls are supported both when answering queries and detecting composite events. To integrate external functionalities in *Retalis*, the corresponding software libraries should be interfaced with *Prolog*.

The interactions between *ELE* and *SLR* are as follows. On the one hand, *ELE* generates composite events. These events constitute the input flow of events to

SLR. *SLR* selectively records these events in its knowledge base. On the other hand, changes in the *SLR* knowledge base trigger corresponding input events for *ELE*. *ELE* can be used, for instance, to detect a pattern of such changes to inform the interested components. In addition, the specification of event patterns of interest in *ELE* can include queries to *SLR*. Queries are used to reason about the domain knowledge and history of events in *SLR*.

An *ELE* program, described in Section 4.1, contains two types of rules. The rules that include the $\leftarrow$ symbol are event rules, specifying patterns of events to derive new events. The rules that include the *:-* symbol are static rules, constituting a *Prolog* program. The specification of the pattern of events in an event rule can include a query to the *Prolog* program defined by the static rules. *Retalis* programs are similar to *ELE* programs. The main difference is that the static rules in *Retalis* are *SLR* rules, constituting a *SLR* program which can be queries from the event rules.

Listing 1 presents an example of how *ELE* and *SLR* are used together in a *Retalis* program. This program records the position of the object segment o_1 whenever the position is changed by more than a meter. This program is read as follows. Capital letters represent variables. The body of the first and third rules are executed when the program is initialized. *c_mem*(m_1,*loc*(o_1,*L*),∞,∞) is a *SLR* clause creating memory m_1 recording the history of *loc*(o_1,*L*)^T events. The second rule is an *ELE* clause querying *SLR*, as written in its *WHERE* clause. For each *seg*(o_1,*C*,*P*,*L*,*PCL*)^T input event, the *prev* clause queries memory container m_1 for the last position of o_1 before time T. If the position has changed by more than a meter, the corresponding *loc*(o_1,*L*) event is generated and recorded in memory m_1. In addition, consumer *moving_objects* is notified by the corresponding event *obj*(o_1)^T. This is specified by the third rule, which is read as follows. The subscription s_1 subscribes consumer *moving_objects* to *loc(O,L)* events with the output template *obj(O)* from time 0. Details of the *ELE* and *SLR* languages are given in Sections 4.1 and 5.1.

```
onProgramStart :- c_mem(m1, loc(o1,L)^T, ∞, ∞).

loc(o1,L)^T <- seg(o1,C,P,L,PCL)^T
                WHERE(
                        prev(m1, loc(o1,L_prev)^T_prev, T)
                        dist(L, L_prev, D),
                        D > 1
                     ).

```

```
onProgramStart :- sub(s1,moving_objects,loc(O,L),obj(O),0).
```

Listing 1: *Retalis* Program Example

A *Retalis* program is parsed and executed by a *Prolog* execution system and is provided a *C++* interface for communication with external components. This makes the *Retalis* language framework-independent, because its core depends only on a *Prolog* execution system. We use *SWI-Prolog*[7] [78] as the *Retalis* execution system and use the *SWI-Prolog C++ interface*[8] to interface the *SWI-Prolog* with *C++*. *Retalis* can be interfaced with existing robotic frameworks mapping its synchronous and asynchronous interfaces to their service-based and publish-subscribe communication mechanisms.

We have developed an interface to integrate *Retalis* with the *ROS* framework [61], the current de-facto standard in open-source robotics. In the *ROS* architecture, each *IEC* is a *ROS* component[9] [61]. Asynchronous and synchronous communications in *ROS* are realized using topics and services, respectively. By subscribing to a topic, a component receives the messages other components publish on that topic. A component invokes a service by sending a request message and receiving a response message.

Figure 3 presents an *IEC* in a *ROS* architecture. *IEC* is subscribed to Topics I_1 and I_2 receiving messages published by the components C_2 and C_3. *IEC* publishes events on topics O_1 and O_2 to which other components are subscribed.

To subscribe an *IEC* to a topic, the *Retalis-ROS* interface requires the name and message type of the topic. This is set in an *XML* configuration file, as in line 4-6 of Listing 2. The *Retalis-ROS* interface offers a number of services to reconfigure the *IEC* at runtime. These include services to subscribe the *IEC* to a topic, to unsubscribe from a topic and to subscribe a topic to events from the *IEC*. To publish an event on a *ROS* topic, the *Retalis-ROS* interface needs to know the message type of that topic. This can be set by the program, as in lines 7-9 of Listing 2, or at runtime.

[7]http://www.swi-prolog.org

[8]http://www.swi-prolog.org/pldoc/package/pl2cpp.html

[9]http://wiki.ROS.org/Nodes

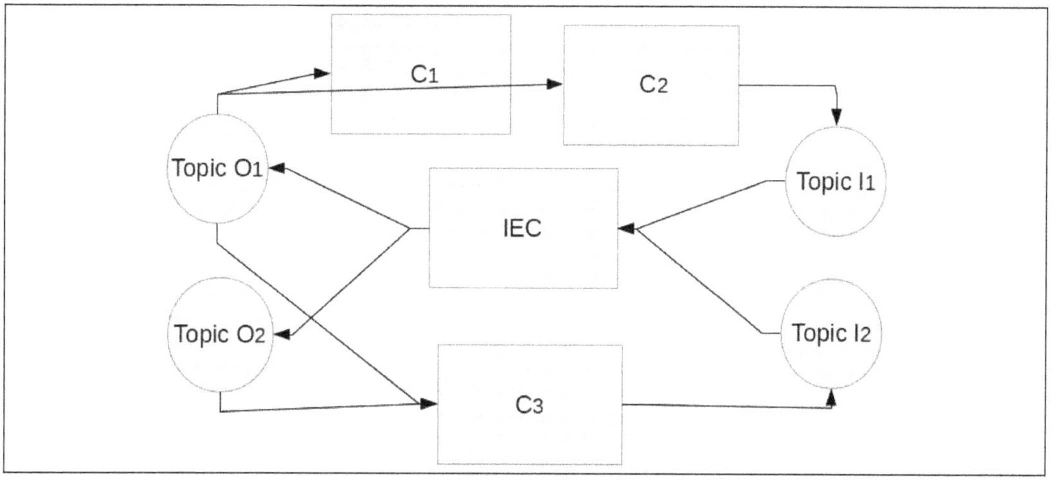

Figure 3: An *IEC* in *ROS* architecture

```
<?XML version="1.0"?>
<publish_subscribe>

  <subscribe_to        name="/ar_pose_marker"
                       msg_type="ar_pose/ARMarkers"
  />
  <publish_to          name="robot_marker_pos"
                       msg_type="geometry_msgs/Transform"
  />
  <publish_to          name="gazeControl"
                       msg_type="headTurn/GazeControl"
  />

</publish_subscribe>
```

Listing 2: *Retalis-ROS XML* configuration file

The conversion among *ROS* messages and *Retalis* events is performed automatically by the *Retalis-ROS* interface. This may be described by an example. Table 1, consisting of five columns, depicts five standard *ROS* message types. The first row in each column is the name of a unique message type. The other rows presents the fields of data that the message type contains. Each field of a message contains a single datum or a list of data, whose type is a basic type such as Integer, Float, String, or it is a *ROS* message type. For example, a *geometry_msgs/Point* message contains three float values and a *geometry_msgs/Pose* message has a *geometry_msgs/Point* message as its first field of data.

geometry_msgs/PoseStamped		std_msgs/Header	
std_msgs/Header	header	uint32	seq
geometry_msgs/Pose	pose	time	stamp
		string	frame_id

geometry_msgs/Pose	
geometry_msgs/Point	p
geometry_msgs/Quaternion	o

geometry_msgs/Point		geometry_msgs/Quaternion	
float64	x	float64	x
float64	y	float64	y
float64	z	float64	z
		float64	w

Table 1: *ROS* message examples

Listing 3 presents the conversion of the *geometry_msgs/PoseStamped ROS* message type to its corresponding *Retalis* event. The conversion maps each *ROS* message to a *Prolog* compound term where the functor symbol of the term is the name of the message type and its arguments are the data fields of the message. Data of basic types such as Integer and Floats are represented by their values. Strings are wrapped by single quotes represented as *Prolog* Strings. Lists of data are represented as *Prolog* lists. Time in *ROS* is a basic data type expressed by two Integer values represented in a *Retalis* event as a list of two numbers.

When converting a *ROS* message to a *Retalis* event, the event is time-stamped with the time-stamp of the header of the message. If the message does not have a header, the event is time-stamped with the system current time. When converting a *Retalis* event to a *ROS* message, the time-stamp of the event is ignored. However, the *Retalis* language provides direct references to time-stamp of events. This can be used to set the *stamp* in the *std_msgs______header(seq,stamp,frame_id)* argument of an event and hence in the header of its corresponding *ROS* message. *ROS* messages from different topics can be of the same type and need to be distinguished. Therefore, we encode topic names as main functor symbols of corresponding *Retalis* events. For example, if the event $p_n(t_1, .., t_n)^z$ is received from the topic x, the event is represented as $x(p_n(t_1, .., t_n))^z$.

```
geometry_msgs_____PoseStamped(
        std_msgs_____Header(seq,stamp,frame_id),
        geometry_msgs_____Pose(
                geometry_msgs_____Point(x,y,z),
                  geometry_msgs_____Quaternion(x,y
                    ,z,w)
                  )
                )^stamp
```

Listing 3: *Retalis* event format corresponding to geometry_msgs/PoseStamped *ROS* message type

4 On-Flow Information Processing

This section discusses on-flow processing requirements of robotic information engineering. It suggests the information flow processing systems [26], and in particular the *ELE* event-processing language [6, 5, 3], as suitable technologies to address the requirements. On-flow processing of data is widespread in large areas of robot software. As examples, we discuss in this section four robotic situations where on-flow processing of data is very useful.

The first situation is decoupling components interacting in robot software. This is usually supported by a publish-subscribe communication mechanism [31] based on an indirect addressing style [20, 80, 61, 42]. The publish-subscribe mechanism organizes robot software in a data-driven manner where components continuously process data generated by the other components. However, due to limited resources of a robot, sensory data needs to be processed selectively. This requires filtering of data passed among components. Data should be filtered based on the robot's operational context, such as its focus of attention. One way to support the filtering of data is to write complex software components whose processes can be reconfigured at runtime. However, such a reconfiguration might not be supported by the available components. The publish-subscribe support in most existing robotic frameworks is limited to topic-based interactions. Providers publish data items on topics, which are received by subscribers to those topics. In these frameworks, a component is usually subscribed to a fixed set of topics. More flexible and context-dependent interaction requires subscribers being able to specify their data of interest based on data patterns and policies [80, 42, 54]. Consider a robot looking for reliable recognition of yellow objects. The object segments sent to the object recognition component should be filtered to include only the yellow and reliably recognized object segments. Another example is the selective processing of new perceptions of object segments by the object recognition component. A new perception of an object should be processed only when the object was perceived at a new location and this location did not change for a given time period.

The second situation is anchoring [24], creating symbolic representation of objects perceived from sensory data. The symbols and the data continuously sensed about the objects should be correlated. In an anchoring process, sensory data is interpreted into a set of hypotheses about recognized objects. For example, in a traffic monitoring scenario [42], images from color and thermal cameras are processed into a set of hypotheses about objects. The object hypotheses need to be correlated over time to deal with the data association problem [11]. There may be false positive and negative percepts, temporal occlusions of objects and visually similar objects in the environment. One can reason also about the hypotheses based on, for instance, the normative characteristics of the physical objects they represent [40, 30]. For example, in the traffic monitoring scenario, one can consider the positions and speeds of objects perceived over time and the layout of the road network. This can be used to reason about stationary and moving objects and their types. For instance, when a car is observed again after a temporary occlusion, it should be assigned the same symbol which was assigned to it previously.

The third situation pertains to flexible plan execution and monitoring in noisy and dynamic environments. The execution of actions/plans are to be driven, monitored and controlled by various conditions [76, 29, 81]. Conditions are monitored by low-level implementations of actions/behaviors to detect their success or failure. However, control and monitoring of plan execution via observation of various conditions at system-level is necessary. The advantages of system level plan execution control and monitoring are to use data provided by different perception components to achieve system's goals, to avoid complicating implementation of actions and to avoid duplicating monitoring functionalities. Depending on an application, conditions to be monitored can be as simple as monitoring an object for being attached to the manipulator. They can be also complex logical, temporal and numerical conditions.

The fourth situation is high-level event recognition to recognize and react in real-time to situations in the environment. One example is detecting traffic violations such as reckless driving by observing qualitative spatial relations among cars [43]. Another example is detecting situations and events such as "successful pass", "successful tackle" and "goal scoring" in football simulation or "washing hand before examination" and "basic clinical examinations carried out in time" in hospital simulations from lower level events [62]. The last example is recognizing human activities such as "cooking", "eating" and "watching TV" in smart homes [58, 66]. Detecting such situations of the environment requires correlating and aggregating sensory data about changes of the environment based on their temporal and logical relations.

What all these situations have in common is a need for processing sensory data flow to extract new knowledge as soon as the relevant data becomes available without requiring persistent storage of data. Supporting on-flow processing requires an expressive and efficient language for real-time processing of data flows based on complex relations among the data items within the flows. On-flow processing is an important requirement in various application domains [26]. In environment monitoring, sensory data is processed to acquire information about the observed world, detect anomalies, or predict disasters. Financial applications analyze stock data to identify trends. Banking fraud detection and network intrusion detection require continuous processing of credit card transactions and network traffic, respectively. *RFID*-based inventory management requires continuous analysis of *RFID* readings. Manufacturing control systems often require observing system behavior to detect anomalies. As the result of many years of research from different research communities on such application domains, a large number of "information flow processing systems" have been developed to support on-flow processing of data [26].

An extensive survey of information flow processing systems [26] shows that the functionalities of these systems are converging to a set of operations and processing policies for on-flow filtering, combining and transformation of data, indicating universal usability of such functionalities for on-flow processing of data. This makes the existing information flow processing systems amenable to support on-flow information processing in robot software.

Current information flow processing research has led to two competing classes of systems [26], Data Stream Management Systems (*DSMS*s) and Complex Event-Processing Systems" (*CEPS*s). *DSMS*s functionalities resemble database management systems. They process generic flow of data through a sequence of transformations based on common *SQL* operators like selections, aggregates and joins. Being an extension of database systems, *DSMS*s focus on producing query answers, which are continuously updated to adapt to the constantly changing contents of their input data. In contrast, *CEPS*s see flowing data items as notification of events happening in the external world. These events should be filtered and combined to detect occurrences of particular patterns of events representing higher level events. *CEPS*s are rooted in publish-subscribe model. They increase the expressive power of subscribing language in traditional publish-subscribe systems with the ability to specify complex event patterns.

Both *DSMS*s and *CEPS*s have their own merits and the recent proposals attempt to combine the best of both classes of systems [26]. However, at this stage, the *CEPS*s are more suitable to support robotic on-flow processing due to the following reasons. First, the semantics given in *CEPS*s to data items as being event notifications naturally corresponds to time-stamped sensory data being observations of the environment by the robot perception components. Second, *CEPSs* put great emphasis on detection and notification of complex patterns of events involving sequence and ordering relations which constitutes a large number of robotic on-flow information engineering problems which is usually out of the scope of *DSMSs*. The rest of this section introduces *ELE*, a state-of-the-art *CEPS*, and discusses its suitability for robotic on-flow information engineering through its comparison with related work.

4.1 *ETALIS* Language for Events (*ELE*)

ELE[10] [6, 5, 3] is an expressive and efficient language with formal declarative semantics for realizing complex event-processing functionalities. *ELE* advances the state-of-the-art *CEPSs* by allowing logical reasoning about domain knowledge in the specification of complex event patterns. Logical reasoning can be used to relate events, accomplish complex filtering and classification of events and enrich events on the fly with relevant background knowledge.

Event-processing functionalities in the *ELE* language are implemented by programming a set of static rules, encoding the domain knowledge and a set of event rules, specifying event patterns of interest to be detected in flow of data. The detected events can themselves match other event patterns, providing a flexible way of composing events in various steps of a hierarchy.

Definition 1 (*ELE* Signature [6]). A signature $\langle C, V, F_n, P^s_n, P^e_n\rangle$ for *ELE* language consists of:
- The set C of constant symbols.
- The set V of variables.
- For $n \in \mathbb{N}$ sets F_n of function symbols of arity n.
- For $n \in \mathbb{N}$ sets P^s_n of static predicate symbols of arity n.
- For $n \in \mathbb{N}$ sets P^e_n of event predicate symbols of arity n with typical elements p^e_n, disjoint from P^s_n.

Based on the *ELE* signature, the following notions are defined.

Definition 2 (Term [6]). A term $t ::= c \mid v \mid f_n(t_1, ..., t_n) \mid p^s_n(t_1, ..., t_n)$.

Definition 3 (Atom [6]). An static/event atom $a ::= p^{s/e}_n(t_1, ...t_n)$ where $p^{s/e}_n$ is a static/event predicate symbol and $t_1, ..., t_n$ are terms.

[10] http://code.google.com/p/etalis/

For example, the *face*(F_i,P_j) event atom is a template for observations of people's faces generated by the *faceRec* component.

Definition 4 (Event [6]). An event is a ground event atom time-stamped with an occurrence time.

- An atomic event refers to an instantaneous occurrence of interest.
- A complex event refers to an occurrence with duration.

For example, the occurrence time of the atomic event *face('Neda',70)*^{28} is time 28 and the occurrence time of the complex event *observed('Neda')*$^{\langle 28,49\rangle}$ is time interval $[28, 49]$.

Definition 5 (*ELE* Rule [6]). An *ELE* rule is a static rule r^s or an event rule r^e.

- A static rule is a Horn clause a :- $a_1, ..., a_n$ where $a, a_1, ..., a_n$ are static atoms. Static rules are used to encode the static knowledge of a domain.
- An event rule is a formula of the type $p^e(t_1, .., t_n) \leftarrow cp$ where cp is an event pattern containing all variables occurring in $p^e(t_1, .., t_n)$. An event rule specifies a complex event to be detected based on a temporal pattern of the occurrence of other events and the static knowledge.

Definition 6 (Event Pattern [6]). The language P of event patterns is

$$P ::= p^e(t_1, ..., t_n) \mid P \;\; WHERE \;\; t \mid q \mid (P).q \mid P \;\; BIN \;\; P \mid not(P).[P, P]$$

where p^e is an n-array event predicate, t_i denote terms, t is a term of type boolean, q is a non-negative rational number, and BIN is one of the binary operators *SEQ, AND, PAR, OR, EQUALS, MEETS, DURING, STARTS*, or *FINISHES*.

4.2 ELE Semantics

As opposed to most *CEPSs*, *ELE* has formal declarative semantics. The input to an *ELE* program is modeled as an event stream, a flow of events. The input event stream specifies that each atomic event occurs at a specific instance of time.

Definition 7 (Event Stream [6]). An event stream $\epsilon : Ground^e \rightarrow 2^{\mathbb{Q}^+}$ is a mapping from ground event atoms to sets of non-negative rational numbers.

For example, $\epsilon(obj(o, c, p)) = \{1, 3\}$ means among all events received by *ELE* as its input over its lifetime, the time points at which the event $objRec(o, c, p)$ occurs are 1 and 3.

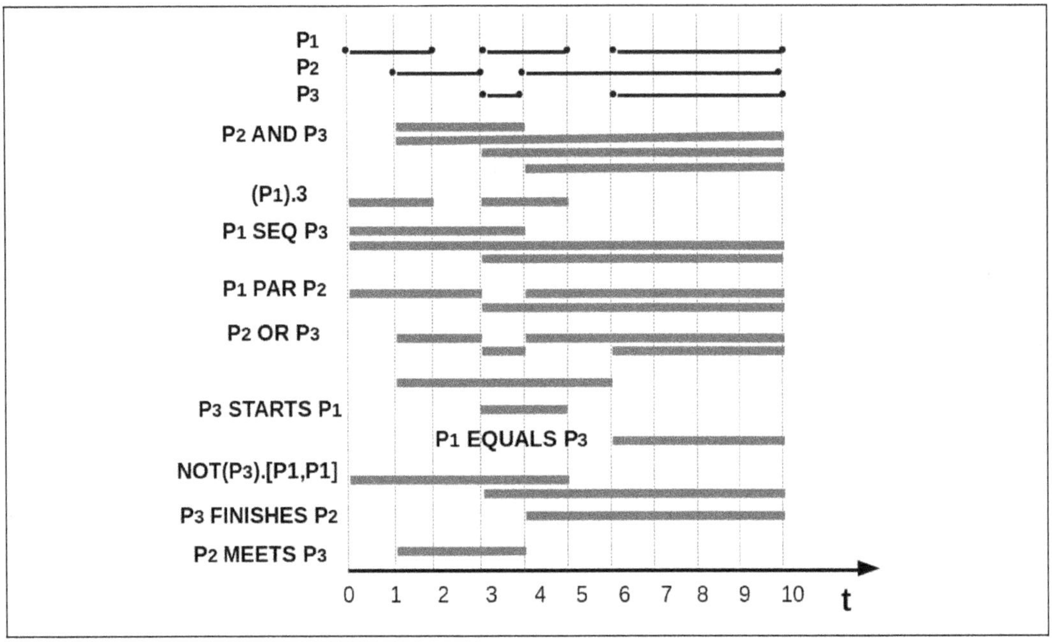

Figure 4: *ELE* event-processing operator examples, re-produced from [6]

Definition 8 (ELE semantics [6]). Given an ELE program with a set R of ELE rules, an event stream ϵ, an event atom a and two non-negative rational numbers q_1 and q_2, the ELE semantics determines whether an event $a^{\langle q_1, q_2 \rangle}$, representing the occurrence of a with the duration $[q1, q2]$, can be inferred from R and ϵ (i.e. $\epsilon, R \models a^{\langle q_1, q_2 \rangle}$).

Figure 4 informally introduces the *ELE* semantics. It provides examples of how *ELE* operators are used to specify complex events in terms of simpler ones. The first three lines show occurrences of the instances of events P_1, P_2 and P_3 during time interval [0,10]. The vertical dashed lines represent units of system time and horizontal bars represent detected complex events for the given patterns. The presented patterns are read as follows:

1. P_2 *AND* P_3: occurrence of both P_2 and P_3.

2. $(P_1).3$: occurrence of P_1 within an interval of length 3 time units.

3. P_1 *SEQ* P_3 : occurrence of P_3 after occurrence of P_1.

4. P_1 *PAR* P_2: occurrence of both P_1 and P_2 with non-zero overlap.

5. P_2 *OR* P_3: occurrence of P_2 or occurrence of P_3

6. P_1 *DURING* (1 *seq* 6): occurrence of P_1 during time interval [1,6]

7. P_3 *STARTS* P_1: occurrence of P_3 and P_1 both starting at the same time and P_3 ending earlier than P_1.

8. P_1 *EQUALS* P_3: occurrence of P_2 and P_3 both at the same time interval

9. $not(P_3).[P_1, P_1]$: occurrence of P_1 after occurrence of another P_1 where there is no occurrence of P_3 in between, during the end of the first P_1 and before the start of the second P_1.

10. P_3 *FINISHES* P_2: occurrence of P_3 and P_2 both ending at the same time and P_3 starting later than P_2.

11. P_2 *MEETS* P_3: occurrence of P_2 and P_3, P_3 starting at the exact time P_2 is ending.

For an example, consider the detection of fire from smoke and high temperature sensor readings. This task is implemented using the following *ELE* rule.

$$
\begin{aligned}
&fireAlam \leftarrow \\
&\qquad smoke(S1)\ AND\ high\_temperature(S2) \\
&\qquad WHERE\ (\ nearby(S1, S2)\ ).
\end{aligned}
$$

This rule is read as follows. *S1* and *S2* are variables. When smoke is detected by a sensor *S1* and high temprature is detected by a sensor *S2*, a fire alarm event is generated, if these sensors are located nearby. If *P2* and *P3* in figure 4 represent smoke and high-temperature events from sensors located nearby, then a fire alarm is generated four times during the time interval [0,10].

The static atom *nearby(S1,S2)* presents an example of logical reasoning in *ELE*. Given an ontology of sensors and their locations, this term specifies whether the sensors are located in the same area. Static atoms can be used to implement arbitrary functionalities in *Prolog*. In addition, they can be used as interface to foreign languages, for instance, to integrate libraries for spatial reasoning. In *Retalis*, *ELE* static terms are replaced by *SLR* queries to, in addition, reason about histories of events.

Complex events are time stamped based on the temporal patterns they represent. For example in Figure 4, the occurrence times of the first instances of *P2* and *P3* events are the intervals [1,3] and [3,4], respectively. According to *ELE* semantics, a fire alarm detected from these events is time stamped with time interval [1,4]. The time stamp of detected patterns can be used to filter the patterns. For example, a fire alarm should be generated, only if both smoke and high temperature are detected within 300 seconds. This condition is added to the fire alarm pattern as follows.

$$
\begin{aligned}
&fireAlam \leftarrow \\
&\qquad (\\
&\qquad smoke(S1)\ AND\ high\_temperature(S2)\\
&\qquad WHERE\ (\ nearby(S1, S2)\ )\\
&\qquad).300.
\end{aligned}
$$

Filters on time intervals of event patterns are important for garbage collection. If the fire alarm pattern does not contain the timing condition, a detection of smoke should be recorded forever in order to generate an alarm whenever a high-temperature is sensed. When the pattern includes the timing condition, the record is deleted after 300 seconds. After this time, the detection of smoke is no longer relevant, even if a high temperature is detected. Irrelevant records of events are automatically deleted by *ELE* garbage collection mechanisms.

ELE is free of operational side-effects, including the order among event rules and delayed or out of order arrival of input events. For example, the sequence pattern in Figure 4 detects three events during the time interval [0,10], no matter the order in which *ELE* receives *P1* and *P3* events.

Listing 4 presents an *ELE* program to illustrate the modeling capabilities of the *ELE* language. In this program, the robot detects an event whenever a person moves an object. Such an event is detected when a person's face is observed while the object is moved.

The program is read piece by piece. The first clause generates a $see(f)$ event for every two immediate consecutive recognitions of a face f, occurring with confidence values over fifty within half a second. The variable F is used to group the recognitions of faces in the event pattern and to pass information to generated events. The rule also explicitly encodes the start and end times of the sequence in content of the generated event by T_s and T_e variables.[11] The second clause detects reliable recognition of objects, when recognized three times within half a second with average confidence value over sixty. *pos_avg* is a static atom computing position of the object by averaging from its perceived positions. The third clause detects cases when an object is moved over five centimetres within a second. The fourth clause combines

[11] This is implemented by adding the $CHECK(t1(T_s), t2(T_e))$ clause which, for brevity, has been omitted.

each two overlapping movement events of an object into a new one with a longer occurrence time. The fifth clause combines two time periods of observing a person if they occur within three seconds after each other. Finally, the last clause detects when an object is moved during the time period a person is being observed.

```
see(F,Ts,Te) <-
        (
         NOT(face(F,P3)).[face(F,P1), face(F,P2)]
         WHERE(P1 > 50, P2 > 50)
        ).0.5s.

relSeg(O,L) <-
        (
         seg(O,C,P1,L1,X1) SEQ seg(O,C,P2,L2,X2)
         SEQ seg(O,C,P3,L3,X3)
         WHERE(pos_avg([L1,L2,L3],L), avg([P1,P2,P3],P), P>60)
        ).0.5s.

mov(O,L1,L2,Ts,Te) <-
        (
         relSeg(O,L1) AND relSeg(O,L2)
         WHERE(dist([L_2,L_1],L), L>0.05)
        ).1s.

mov(O,L1,L4,T1,T4) <-
        mov(O,L1,L2,T1,T2) PAR mov(O,L3,L4,T3,T4)
        WHERE(T3>T1).

see(F,T1,T4) <-
         (see(F,T1,T2) SEQ see(F,T3,T4))
         OR
         (see(F,T1,T2) MEETS see(F,T3,T4))
         WHERE(T3-T2<3).

movBy(O,F,L2,T2) <-
         mov(O,L1,L2,T1,T2) DURING see(F,T1,T2).
```

Listing 4: An *ELE* program for monitoring objects moved by humans

Assume an object has moved while the robot was seeing a face of a person. If the robot continues to see the face, the above rules generate more and more events indicating the person has moved the object, but one of such events might be sufficient for an application. Each time a new event occurs, the event along with the past events can match the pattern of a rule in several ways.

The *ELE* language offers various *consumption policies* to filter our repetitive rule firings. These includes policies to select a particular pattern among possible matches and to limit the use of an event to fire a rule more than once. While such policies are not aligned with declarative semantics of *ELE*, they are widely adopted in *CEPS*s for practical reasons. *ELE* also supports adding or deleting *ELE* rules at runtime allowing flexible reconfiguration of event-processing functionalities.

4.3 Runtime Subscription in *Retalis*

The *ELE* interface facilitates programming a fixed set of output channels to deliver certain types of events to consumers. *Retalis* extends this functionality enabling robot software components to subscribe to *Retalis* for their events of interest at run-time. The events are sent to subscribers asynchronously as soon as they are processed by *Retalis*.

A component subscribes to *Retalis* by sending a subscription request using a *ROS* service that the *Retalis* interface provides. A subscription is of the form $subscribe(Topic, Q, Tmpl, T_s, T_e)$. The process of the request by *Retalis* results in subscribing *Topic* to events matching the query pattern Q that have occurred during time interval $[T_s, T_e]$. A query pattern Q is a tuple $\langle e, Cond \rangle$, where e is an event atom and $Cond$ is a set of conditions on variables which are arguments of e. An event P matches a query pattern Q when there is a substitution which can unify p and e and makes the conditions in *Cond* true (i.e. $\exists\theta(p = q_\theta)$).

When a subscription is registered, every event matching the subscription is asynchronously sent to the corresponding topic as the event is read from the *Retalis* input or generated by *ELE* rules. Events are first converted to the template form *Tmpl* before being sent to the topic. If a component does not know in advance the end time of its subscription, it can subscribe to its events of interest using $sub(Id, C, Q, Tmpl, T_s)$ and unsubscribe from them at any time using $unsub(Id, T_e)$. Id is a unique identifier of such a subscription.

Example 1. When the robot is asked to follow the object segment $seg11$, the control component sets the target location for the Gaze component to the location of $seg11$ by sending the following subscription command to the Information-

Engineering Component.

$$sub(100, \text{`}camCtrl\text{'}, \langle relObj(\text{`}seg11\text{'}, L), \langle\rangle\rangle, pos\_goal(L), \text{`}now\text{'})$$

Consequently, every time *IEC* processes an event $relObj(\text{`}seg11\text{'}, L)$, it sends the location $L$ of $seg11$ to the Gaze in the $pos\_goal(L)$ format. To unsubscribe, the control component sends the $unsub(100, \text{`}now\text{'})$ command to *IEC*.

4.4 Discussion

Previous robotic research is concerned with on-flow processing for specific research tasks such as component interaction, anchoring, monitoring and event-recognition. The consequence is the narrow scope of related robotic research reducing the community collaboration in supporting on-flow processing in robotic software. For instance, on-flow processing support of open-source robotic software such as *ROS* is limited to fixed publish-subscribe flow of data among components.

In parallel to this research, the *DyKnow* [42, 38] framework has been extended with a number of tools that are relevant to on-flow processing [27, 44, 39]. The main feature of the work is the annotation of data streams and transformation processes with semantic descriptions. The semantic descriptions are used for automatic construction of streams of data. The *C-SPARQL* [12] language has been integrated to support the querying of flows of data. *C-SPARQL* belongs to the *DSMS*s category of on-flow processing systems. The advantages of *ELE* over *C-SPARQL* is its support for capturing complex data patterns. In contrast, the *Retalis* does not support an automatic discovery of flows of data, for instance, required to detect a complex event. The input and output subscriptions of Information-Engineering Components and the event patterns they process are reconfigurable at runtime. However, such reconfigurations are not made automatic.

The literature does not contain a comparison between the expressive power of information flow processing systems. *ELE* is one of the most expressive systems as it supports most of the existing information flow processing operations listed in [26]. In particular, *ELE* supports the representation of all possible thirteen temporal relations between time interval occurrence times of two events as defined in Allen's interval algebra [2], non-occurrence of an event between the occurrence of two other events, and iterative and aggregating patterns. Furthermore, arbitrary processes can be applied on events through the use of static atoms in *ELE* syntax, provided that such processes are interfaced with the *Prolog* language. An example is interfacing spatial reasoning functionalities with *Prolog* presented in [72].

Logic-based approaches such as *Chronicle Recognition* [33] and *Event Calculus* [47, 67] have received considerable attention for event representation and recognition due to their merits, including expressiveness, formal and declarative semantics and being supported by machine learning tools to automate the construction and refinement of event recognition rules [8, 6]. However, the query-response execution mode and scalability of classic logic-based systems limits their usability for on-flow information processing. The query-response execution means detecting an event at runtime requires frequently querying the system for that event. Moreover, the event is detected only when the next time the system is queried for that event. In addition, efficient evaluation of such queries requires caching mechanisms not to re-evaluate queries over all historic data [22]. *ELE* bridges the gap between *CEPS*s and logic-based event-recognition systems by offering a logic-based *CEPS* with an event-driven, incremental and efficient execution model.

The *IDA* [80, 55] and *CAST* [37, 36] are robotic frameworks supporting the subscription of components to their events of interest based on the type and content of events. Using *XML* data format in *IDA*, a subscriber can register for information items containing specific field of data. *IDA* also provides few types of event filters such as the *Frequency filter*, which outputs only every n-th received notification. *Retalis* provides a general framework to address a much wider variety of event processing requirements, including temporal and spatial reasoning over events to detect complex event patterns. Moreover, the subscription mechanisms of *IDA* and *CAST* are tightly built over their underlying middleware. In contrast, *Retalis* is framework-independent and has been interfaced with *ROS* which is widely used by robotic community.

The use of *CEPS*s for detecting high-level events in agent research has been proposed before. Buford et al. [22] extend the *BDI* architecture with situation management components for event correlation in distributed large-scale systems. Ranathunga et al. [23] utilize the *ESPER*[12] event-processing language to detect high-level events in second life virtual environments.[13] However this work is not concerned with the robotic on-flow information-processing problem, it does not provide a formal account of event processing and does not support run-time subscription. Other related work includes various approaches for high-level event recognition, anchoring and monitoring, for instance, using Chronicle recognition, constraint satisfaction or variants of temporal logic [42, 58, 43, 28]. Such approaches do not satisfy all on-flow information processing requirements. For instance, the Chronicle recognition or constrained satisfaction approaches based on simple temporal networks cannot

[12]Esper Reference, Esper Team and EsperTech Inc, accessible at http://esper.codehaus.org/esper-4.9.0/doc/reference/en-US/html_single/

[13]http://secondlife.com

express atemporal constraints, and temporal logic based approaches do not support transformation of information.

5 On-Demand Information Processing

On-demand information processing corresponds to managing data in memory or knowledge base to be queried and reasoned upon on request. This section presents the *SLR* language[14] to address on-demand processing requirements related to discreteness, asynchronicity and continuity of robotic sensory data that are not satisfactorily supported by existing systems. After a short introduction of these requirements, the *SLR* syntax and semantics are presented and the usability of the language and its relation with existing works is discussed.

Building robot knowledge based on discrete observations is not always a straightforward task, since events contain various information types that should be represented and treated differently. For example, to accurately calculate the robot position at a time point, one needs to interpolate its value based on the discrete observations of its value in time. One also needs to deal with the persistence of knowledge and its temporal validity. For example, it might be reasonable to assume that the color of an object remains the same until a new observation is made indicating the change of color. In some other cases, it may not be safe to infer an information, such as the location of an object, based on an observation that is made in distant past. Building robot knowledge of its environment upon sensory events requires language support to simplify reasoning about the state of the environment at a time based on discrete observations of the environment.

A network of distributed and parallel components process robot sensory data and send the resulting events to the knowledge base. Due to processing times of the perception components and possible network delay, the knowledge base may receive the events with some delays and not necessarily in the order of their occurrence. For example, the event indicating the recognition of an object in a *3D* image is generated by the object recognition component sometime after the actual time at which the object is observed, because of object recognition processing time. Another example is when data is generated or needs to be verified by an external source with arbitrary operating time. Therefore, when the knowledge base is queried, correct evaluation of the query may require waiting for the perception components to finish processing of sensory data to ensure that all data necessary to evaluate the query is present in the knowledge base. For example, the query, "how many cups are on the table at time t?" should not be answered immediately at time t, but answering the query

[14] An earlier version of *SLR* is appeared in a technical report before [83].

should be delayed until after completing the processing of pictures of the table by the object recognition component and the reception of the results by the knowledge base. Dealing with asynchronicity of sensory data requires supporting the implementation of synchronization mechanisms to assure evaluating queries when relevant data to queries are available in the knowledge base.

Robot perception components continuously send their observations to the knowledge base, leading to a growth of memory required to store and maintain the robot knowledge. The unlimited growth of the event history leads to a degradation of the efficiency of query evaluation and may even lead to memory exhaustion. Bounding the growth of memory requires supporting the implementation of mechanisms to prune outdated data.

5.1 *SLR* Language for Event Management and Querying

Synchronized Logical Reasoning language (*SLR*) is a knowledge management and querying language for robotic software enabling the high-level representation, querying and maintenance of robot knowledge. In particular, *SLR* aims at simplifying the representation of robot knowledge based on its discrete and asynchronous observations and improving efficiency and accuracy of query evaluation by providing synchronization and event-history management mechanisms. These mechanisms facilitate ensuring that all data necessary to answer a query is gathered before the query is answered and that outdated and unnecessary data is removed from memory.

In an Information-Engineering Component programmed in *Retalis*, the input to *SLR* is the stream of events processed by *ELE*. This consists of the input stream of events to the *IEC*, time-stamped by the perception components and the events generated and time-stamped by *ELE*. The *SLR* language bears close resemblance to logic programming and is both in syntax and semantics very similar to *Prolog*. Therefore, we first review the main elements of *Prolog* upon which we define the *SLR* language.

In *Prolog* syntax, a *term* is an expression of the form $p(t_1, \ldots, t_n)$, where p is a functor symbol and $t_1, \ldots, t_n$ are constants, variables or terms. A term is *ground* if it contains no variables. A *Horn clause* is of the form $a_1 \wedge \ldots \wedge a_n \rightarrow a$, where a is a term called the *Head* of the clause, and $a_1, \ldots, a_n$ is called the *Body* where a_i are terms or negation of terms. $a \leftarrow true$ is called a *fact* and usually written as a. A *Prolog program* P is a finite set of *Horn clauses*.

One executes a logic program by asking it a query. *Prolog* employs the *SLDNF* resolution method [7] to determine whether or not a query follows from the program. Given a goal, *SLDNF* tries to prove the goal using the rules and facts of the program.

A goal is proved if there is a variable substitution by applying which the goal matches a fact, or matches the head of a rule and the goals in body of the rule can be proved from left to right. Goals are resolved by trying the facts and rules in the order they appear in the program. A query may result in a substitution of free variables. We use $P \vdash_{SLDNF} Q\theta$ to denote a query Q on a program P, resulting in a substitution θ.

5.1.1 *SLR* Syntax

An *SLR* signature includes constant symbols, *Floating-point* numbers, variables, time points, and two types of functor symbols. Some functor symbols are ordinary *Prolog* functor symbols called *static functor symbols*, while the others are called *event functor symbols*.

Definition 9 (*SLR* Signature). A signature $S = \langle C, R, V, Z, P^s, P^e \rangle$ for *SLR* language consists of:
- A set C of *constant symbols.*
- A set $R \subseteq \mathbb{R}$ of *real numbers.*
- A set V of *variables.*
- A set $Z \subseteq R_{r \geq 0} \cup V$ of *time points*
- P^s, a set of P^s_n of *static functor symbols* of arity n for $n \in \mathbb{N}$.
- P^e, a set of P^e_n of *event functor symbols* of arity n for $n \in \mathbb{N}_{n \geq 2}$, disjoint with P^s_n.

Definition 10 (Term). A *static/event term* is of the form $t ::= p^s_n(t_1, ..., t_n)/p^e_n(t_1, ..., t_{n-2}, z_1, z_2)$ where $p^s_n \in P^s_n$ and $p^e_n \in P^e_n$ are *static/event functor symbols*, t_i are *constant symbols*, *real numbers*, *variables* or *terms* themselves and z_1, z_2 are *time points* such that $z_1 \leq z_2$.

For the sake of readability, an event term is denoted as $p_n(t_1, \ldots, t_{n-2})^{[z_1,z_2]}$. Moreover, an event term whose z_1 and z_2 are identical is denoted as $p_n(t_1, \ldots, t_{n-2})^z$.

Definition 11 (Event). An *event* is a ground event term $p_n(t_1, ...t_n)^{[z_1,z_2]}$, where z_1 is called the *start time* of the event and z_2 is called its *end time.* The functor symbol p_n of an event is called its *event type.*[15]

We introduce two types of static terms, *next* and *prev* which respectively refer to occurrence of an event of a certain type observed right after and right before a time

[15]The representation of events in *SLR* and *ELE* is similar, but the *SLR* signature is defined in a way to be close to *Prolog.*

point, if such an event exists. In the next section we provide the semantics. In this section, we restrict ourselves to the syntax of *SLR*.

Definition 12 (Next Term). Given a signature S, a next term of the form $next(p_n(t_1, ...t_n)^{[z_1,z_2]}, z_s, z_e)$ has an $p_n(t_1, ...t_n)^{[z_1,z_2]}$ event term and two time points z_s, z_e representing a time interval $[z_s, z_e]$ as its arguments.

Definition 13 (Previous Term). Given a signature S, a previous term of the form $prev(p_n(t_1, ...t_n)^{[z_1,z_2]}, z_s)$ has an event term $p_n(t_1, ...t_n)^{[z_1,z_2]}$ and a time point z_s as its arguments.

Definition 14 (*SLR* Program). Given a signature S, an *SLR* program D consists of a finite set of Horn clauses of the form $a_1 \wedge \ldots \wedge a_n \rightarrow a$ built from the signature S, where *next* and *prev* terms can only appear in the body of rules and the program excludes event facts (i.e. events).

5.1.2 *SLR* Operational Semantics

An *SLR* knowledge base is modeled as an *SLR* program and an input stream of events. In order to limit the scope of queries on a *SLR* knowledge base, we introduce a notion of an event stream view, which contains all events occurring up to a certain time point.

Definition 15 (Event Stream). An *event stream* ϵ is a (possibly infinite) set of events.

Definition 16 (Event Stream View). An *event stream view* $\epsilon(z)$ is the maximum subset of event stream ϵ such that events in $\epsilon(z)$ have their end time before or at time point z, i.e. $\epsilon(z) = \{p_n(t_1, \ldots, t_{n-2})^{[z_1,z_2]} \in \epsilon \mid z_2 \leq z\}$.

Definition 17 (Knowledge Base). Given a signature S, a knowledge base k is a tuple $\langle D, \epsilon \rangle$ where D is an *SLR* program and ϵ is an event stream defined upon S.

Definition 18 (*SLR* Query). Given a signature S, an *SLR* query $\langle Q, z \rangle$ on an *SLR* knowledge base k consists of a regular *Prolog* query Q built from the signature S and a time point z. We write $k \vdash_{SLR} \langle Q, z \rangle \theta$ to denote an *SLR* query $\langle Q, z \rangle$ on a knowledge base k, resulting in a substitution θ.

The operational semantics of *SLR* for query evaluation follows the standard *Prolog* operational semantics (i.e. unification, resolution and backtracking) [7] as follows: The evaluation of a query $\langle Q, z \rangle$ given an *SLR* knowledge base $k = \langle D, \epsilon \rangle$

consists in performing a depth-first search to find a variable binding that enables derivation of Q from the rules and static facts in D, and events in ϵ. The result is a set of substitutions (i.e. variable bindings) θ such that $D \cup \epsilon \vdash_{SLDNF} Q\theta$ under the condition that event terms which are not arguments of *next* and *prev* terms can be unified with events that belonging to $\epsilon(z)$.

The event stream models observations made by robot perception components. Events are added to the *SLR* knowledge base in the form of facts when new observations are made. The z parameter of a query sets the scope of the query to set of observations made up until time z. This means that the query $\langle Q, z \rangle$ cannot be evaluated before time z, since *SLR* would not have received the robot's observations necessary to evaluate Q and the query can be evaluated as soon as all observations up to time z is in place. The only exceptions are the *prev* and *next* clauses whose evaluation might need observations made after time z. A query $\langle Q, z \rangle$ can be posted to *SLR* long after time z, in which case the *SLR* knowledge base contains observations made after time z. In order to have a clear semantics of queries, *SLR* evaluates a query $\langle Q, z \rangle$ by only taking into account the event facts in $\epsilon(z)$. Regardless of the z parameters of queries, the *next* or *prev* clauses are evaluated based on their declarative definitions as follows.

Definition 19 (Previous Term Semantics). A $prev(p_n(t_1, ...t_n)^{[z_1,z_2]}, z_s)$ term unifies $p_n(t_1, ...t_n)^{[z_1,z_2]}$ with an event $p_n(t'_1, ...t'_n)^{[z'_1,z'_2]}$ in $\epsilon(z_s)$ such that there is no other such event in $\epsilon(z_s)$ that has its end time later than z'_2. If such a unification is found, the *prev* clause succeeds and fails otherwise.

$$prev(p_n(t_1, ...t_n)^{[z_1,z_2]}, z_s) : \begin{cases} \theta & \exists p_n(t'_1, ...t'_n)^{[z'_1,z'_2]} \in \epsilon(z_s) | \\ & \quad \exists \theta((p_n(t_1, ...t_n)^{[z_1,z_2]})_\theta = (p_n(t'_1, ...t'_n)^{[z'_1,z'_2]})_\theta) \\ & \quad \wedge \nexists p_n(t"_1, ...t"_n)^{[z"_1,z"_2]} \in \epsilon(z_s) | \\ & \quad\quad z"_2 > z'_2 \wedge \\ & \quad\quad \exists \gamma((p_n(t_1, ...t_n)^{[z_1,z_2]}) \stackrel{\gamma}{=} \\ & \quad\quad\quad (p_n(t"_1, ...t"_n)^{[z"_1,z"_2]})), \\ \text{fails} & \text{otherwise} \end{cases}$$

By definition, the variable z_s should be already instantiated when a *prev* clause is evaluated and an error is generated otherwise. It is also worth noting that a *prev* clause can be evaluated only after time z_s when all relevant events with end time earlier or equal to z_s have been received by and stored in the *SLR* knowledge base.

Definition 20 (Next Term Semantics). A $next(p_n(t_1, ...t_n)^{[z_1,z_2]}, z_s, z_e)$ term unifies $p_n(t_1, ...t_n)^{[z_1,z_2]}$ with an event $p_n(t'_1, ...t'_n)^{[z'_1,z'_2]}$ in $\epsilon(z_e)$ such that $z_s \leq z'_2 \leq z_e$ and there is no other such event in ϵ that has its end time earlier than z'_2. If such a unification is found, the *next* clause succeeds and fails otherwise.

$$next(p_n(t_1,...t_n)^{[z_1,z_2]}, z_s, z_e): \begin{cases} \theta & \exists p_n(t'_1,...t'_n)^{[z'_1,z'_2]} \in \epsilon(z_e)| \\ & \quad z'_2 \geq z_s \wedge \\ & \quad \exists\theta((p_n(t_1,...t_n)^{[z_1,z_2]})_\theta = (p_n(t'_1,...t'_n)^{[z'_1,z'_2]})_\theta) \\ & \quad \wedge \nexists p_n(t"_1,...t"_n)^{[z"_1,z"_2]} \in \epsilon(z_e)| \\ & \quad z_s \leq z"_2 < z'_2 \wedge \\ & \quad \exists\gamma((p_n(t_1,...t_n)^{[z_1,z_2]}) \stackrel{\gamma}{=} \\ & \qquad (p_n(t"_1,...t"_n)^{[z"_1,z"_2]})), \\ \text{fails} & \text{otherwise} \end{cases}$$

By definition, the variables z_s and z_e should be instantiated when a *next* clause is evaluated and an error is generated otherwise. A *next* clause can only be evaluated after time z_e when all relevant events with end time earlier or equal to z_e have been received and stored in the *SLR* knowledge base. However, if we assume that events of the same type (i.e. with same functor symbol and arity) are received by *SLR* in the order of their end times, the next clause can be evaluated as soon as *SLR* receives the first event with the end time equal or later than z_s which is unifiable with $p_n(t_1,...t_n)^{[z_1,z_2]}$, not to unnecessarily postpone queries.

The *next* and *prev* clauses can be implemented by the following two *Prolog* rules in which the $\neg$ symbol represents *Negation as failure.* However, we take advantage of the fact that *SLR* usually receives events of the same type in the order of their end times. *SLR* maintains the sorted list of events of each type ordered by their end times whose maintenance usually only requiring the assertion of events by the *asserta Prolog* built-in predicate. In this way, finding a previous/next event of a type occurring before/after a time point requires examining only a part of the history of those events.

$$\begin{aligned} prev(p_n(t_1,...t_n)^{[z_1,z_2]}, z_s)\text{:-}&p_n(t_1,...t_n)^{[z_1,z_2]}, z_2 \leq z_s, \\ &\neg(p_n(t_1",...t_n")^{[z_1",z_2"]},\ z_2" \leq z_s, z_2" > z_2). \quad (1) \\ next(p_n(t_1,...t_n)^{[z_1,z_2]}, z_s, z_e)\text{:-}&p_n(t_1,...t_n)^{[z_1,z_2]},\ z_s \leq z_2 \leq z_e, \\ &\neg(p_n(t_1",...t_n")^{[z_1",z_2"]},\ z_s \leq z_2" \leq z_e, z_2" < z_2). \quad (2) \end{aligned}$$

5.1.3 State-Based Knowledge Representation

SLR aims at simplifying the transformation of events into a state-based representation of knowledge, using derived facts. The following paragraphs presents some typical cases where a state-based representation is more suitable and how it is realized in *SLR*.

Persistent Knowledge Persistent knowledge refers to information that is assumed not to change over time.

Example 2. The following rule specifies that the color of an object at a time T is the color that the object was perceived to have at its last observation.

$$color(O, C)^T \text{:- } prev(obj(O,, C)^Z, T). \tag{3}$$

Persistence with Temporal Validity The temporal validity of persistence refers to the period when it is assumed that information derived from an observation remains valid.

Example 3. To pick up an object O, its location should be determined and sent to a planner to produce a trajectory for the manipulator to perform the action. This task can be naively presented as the sequence of actions: determine the object's location L, compute a manipulation trajectory *Trj*, and perform the manipulation. However, due to environment dynamics and interleaving in task execution, the robot needs to check that the object's location has not been changed and the computed trajectory is still valid before executing the actual manipulation task. The following three rules can be used to determine the location of an object and its validity as follows. If the last observation of the object is within the last five seconds, the object location is set to the location at which the object was seen last time. If the last observation was made longer than five seconds ago, the second rule specifies that the location is outdated. The third rule sets the location to "never-observed", if the robot has never observed such an object. The symbol ! represents *Prolog* cut operator and locations are assumed to be absolute.

$$location(O, L)^T \text{:- } prev(seg(O, L)^Z, T), T - Z \leq 5, !. \tag{4}$$

$$location(O, \text{"}outdated\text{"})^T \text{:- } prev(seg(O, L)^Z, T),\ T - Z > 5, !. \tag{5}$$

$$location(O, \text{"}never\text{-}observed\text{"})^T. \tag{6}$$

Continuous Knowledge Continuous knowledge refers to information from a continuous domain.

Example 4. The following rule calculates the camera to base relative position L at a time T. It interpolates from the last observation L_1 before T to the first observation L_2 after T. *est* is a user defined term performing the actual interpolation.

$$tf(cam, base, L)^T \text{:- } prev(tf(cam, base, L_1)^{T_1}, T), \\ next(tf(cam, base, L_2)^{T_2}, [T, \infty]), est([L, T], [L_1, T_1], [L_2, T_2]). \tag{7}$$

The following rule similarly interpolates the base to world relative position L at a time T. However, if the position is not observed within a second after time T, the position is assumed without change and is set to its last observed value. The $\rightarrow$ symbol represents *Prolog* "If-Then-Else" choice operator.

$$tf(base, rcf, L)^T \text{ :- } prev(tf(base, rcf, L_1)^{T_1}, T),$$
$$(next(tf(base, rcf, L_2)^{T_2}, T, T+1) \rightarrow est([L, T], [L_1, L_2], [L_2, T_2]) \ ; \ L \ is \ L_1). \quad (8)$$

The following *ELE* rule concerns recognition of an object O at a position L_{o-c} relative to the camera at a time T. It generates a corresponding *segR* event. It calculates the object position in the reference coordination frame by querying the *SLR* knowledge base. The camera to base and base to world relative positions at time T are estimated by rules (7) and (8).

$$segR(O, L) \leftarrow seg(O, L_{o-c})^T \ WHERE(\ tf(cam, base, L_{c-b})^T,$$
$$tf(base, rcf, L_{b-rcf})^T,$$
$$mul([L_{o-c}, L_{c-b}, L_{b-rcf}], L) \ ). \quad (9)$$

5.1.4 Active Memory

SLR supports selective recording and maintenance of data in knowledge bases using memory instances.

Definition 21 (Memory Instance). A memory instance with an id Id, a query Q and a policy $\langle L, N\rangle$ keeps the record of a subset of input events to *SLR*: the events that match the query Q such that at each time T, the memory instance only contains the events which have their end times within the last L seconds and only includes the recent N number of such events ordered by their end time. An id is a ground term and a query is of the form $\langle e, Cond\rangle$, where e is an event atom and $Cond$ is a set of conditions on variables that are arguments of e. An event P matches a query pattern Q when there is a substitution that can unify p and e and makes the conditions in $Cond$ true (i.e. $\exists\theta(p = q_\theta)$).

Memory instances are created by executing queries of the form *c_mem(Id,Q,N,L)* on the *SLR* knowledge base in initialization of the *SLR* program. They can also be created at runtime by *ELE* rules or by external components using a *ROS* service the *IEC* provides. Similarly, memory instances are deleted at runtime by executing queries of the form *d_mem*(Id) each deleting all memory instances whose id_i match the term Id (i.e. $\exists\theta(id = Id_\theta)$).

Example 5. The $c\_mem(tf, \langle tf(X,Y,Z), \langle\rangle\rangle, \infty, 300)$ query creates a memory instance to keep the history of $tf(X,Y,Z)^T$ events from the *stateRec* component for 300 seconds. In the rule (9), we saw that the *SLR* knowledge base is queried to position object segments in the reference coordination frame. If we assume that the *IEC* receives data of object segments within 300 seconds since they appear in front of the camera, then we only need to keep the history of *tf* events for 300 seconds. In another example, for each object o_i in *segR(O,L)* events, the *ELE* rule (10) generates a memory instance with the corresponding id of $obj(o_i)$. A memory instance is generated, if it does not already exist. This is checked using the $\neg exist\_mem(monitor(O))$ clause. Each memory instance $obj(o_i)$ keeps the last occurrence of $segR(o_i, L)$ events at which o_i is located on the floor, checked by the *onFloor Prolog* term implementing the required spatial inference. The use of *DO* clause is another way of performing *SLR* queries in *ELE* syntax.

$$Do(c\_mem(obj(O), \langle segR(X,L), \langle X == O, onFloor(L)\rangle\rangle, 1, \infty)) \leftarrow segR(O,L) WHERE(\neg exist\_mem(obj(O))). \quad (10)$$

The histories of events maintained in memory instances are accessed in the *SLR* program using the following static terms.

Definition 22 (Memory Term Semantics). A $mem(Id, X)$ term unifies X with an event $p_n(t_1,..,t_{n-2})^{[z1,z2]}$ that belongs to a memory instance whose id matches the term Id (i.e. $\exists\theta(id = Id_\theta)$). When backtracking over a $mem(Id, X)$ term in evaluating an *SLR* query, the possible unification of X is checked against all events recorded in all such memory instances.

Definition 23 (Previous_Memory Term Semantics). A term of the form $prev(Id, X, Z_s)$, where Id is a ground term, unifies X with an event which has the latest occurrence time among the events that belong to the memory instance Id, are unifiable with X and have their end time before or equal to Z_s. The term fails if such a unification is not found.

Definition 24 (Next_Memory Term Semantics). A $next(Id, X, z_s, z_e)$ term, where Id is a ground term, unifies X with an event which has the earliest occurrence time among the events that belong to the memory instance Id, are unifiable with X and have their end time within time interval $[Z_s, Z_e]$. The term fails if such a unification is not found.

Example 6. The rule (11) re-writes the rule (8) by querying the previous event of the form *tf(base,rcf,L)* occurring before T and the next *tf(base,rcf,L)* event occurring during $[T, T+1]$ from the memory instance *tf*, defined in the previous example to keep the history of *tf* events for 300 seconds. Another example is the query *findAll*(X, *mem(obj(O),X)*, *List*) which queries all $obj(O)$ memory instances created by the rule (10) for their records of $segR(X, L)$ events using the $mem(obj(O), X)$ template and put the list of results in the variable *List*.

$$\begin{aligned} tf(base, rcf, L)^T \text{ :- } & prev(tf, tf(base, rcf, L_1)^{T_1}, T), \\ & (next(tf, tf(base, rcf, L_2)^{T_2}, T, T+1) \rightarrow \\ & est([L, T], [L_1, L_2], [L_2, T_2]) \text{ ; } L \text{ } is \text{ } L_1). \end{aligned} \quad (11)$$

SLR generates events when memory instances are created, deleted or updated. Memory events are fed to *ELE* as input. Consequently, patterns of memory events can be captured by *ELE* to notify external components with information about changes of memory. Memory events are also used internally to keep track of the latest update time of memory instances. This mechanism is used to synchronized queries, discussed in Section 5.1.5.

This mechanism can be used to generate all sorts of events related to changes of the memory such as the addition or deletion of memory instances or even the addition or deletion of events to/from memory instances.

5.1.5 Synchronizing Queries over Asynchronous Events

SLR supports the synchronization of queries to deal with the delayed and out of order reception of sensory data to the knowledge base.

Definition 25 (Event Process Time). The process time (i.e. $t_p(e)$) of an event e is the time at which the event is received by and added to the *SLR* knowledge base (i.e. processed by *IEC*).

Definition 26 (Event Delay Time). The delay time ($t_d(e)$) of an event e is the difference between its process time and its end time (i.e. $t_d(p^{[z1,z2]}) = t_p(p^{[z1,z2]}) - z2$).

A query should be evaluated after all events relevant to the query have been already received by the *SLR* knowledge base. The parameter z of a query $\langle goal, z \rangle$ limits the scope of the query to observations made up until time z. To evaluate the *goal*, a number of memory instances are queried. Therefore, all relevant events to these memory instances occurring up to time z should have been received by *SLR* before performing the query.

Definition 27 (History Availability). The history of events of a type p_n up to a time z is available at a time t when at this time the *SLR* has received all events of type p_n occurring by time z (having end time earlier or equal to z).

Moreover, all previous and next memory terms should be correctly evaluated according to their definitions. Finding the previous event of type $p_n(t_1, .., t_n)$ occurring up to time z_s requires having received all $p_n(t_1, .., t_n)$ events occurring up until time z_s. If we assume events of each type are received by *SLR* in the order of their end times, then finding the next event of type $p_n(t_1, .., t_n)$ occurring within time interval $[z_s, z_e]$ requires having received the first $p_n(t_1, .., t_n)$ event which has its end time equal or more than z_s, or make sure that no $p_n(t_1, .., t_n)$ event has occurred during $[z_s, z_e]$. *SLR* postpones an individual query[16] when necessary until it is achievable, as defined below.

Definition 28 (Dynamic Goal Set of Query). The dynamic goal set of a query $\langle goal, z\rangle$ for an *SLR* program D is the set of all $mem(Id, X)$, $prev(Id, Z_s)$ and $next(Id, Z_s, Z_e)$ predicates that can possibly be queried when evaluating the *goal* on the knowledge base. The dynamic goal set can be determined by going through all rules in D using which the *goal* could be possibly proven and gathering all $mem(Id, X)$, $prev(Id, Z_s)$ and $next(Id, Z_s, Z_e)$ terms appearing in bodies of those rules.

Definition 29 (Query Achievability). A query $\langle goal, z\rangle$ is achievable when three conditions are met. First, the histories of all relevant events to memory instances in dynamic goal set of the query are available up to time z. Second, for each $prev(Id, Z_s)$ term in the dynamic goal set of the query, the history of all relevant events up to time Z_s is available. Third, for each $next(Id, Z_s, Z_e)$ term in the dynamic goal set of the query, a relevant event has been received or the history of all relevant events up to time z_e is available.

To determine when the history of events of a type p_n up to a time z is available, *SLR* can be programmed in two complementary ways. One way is to set a maximum delay time (i.e. $t_{d_{max}}$) for events of each type. When the system time passes $t_{d_{max}}(p_n)$ seconds after z, *SLR* assumes that the history of events of type p_n up to time z is available. The maximum delay times of events depends on the runtime of the components generating them and need to be approximated. The maximum delay times can be set the system developer. It can also be approximated by *SLR* as follows. Whenever an event of type p_n is processed, *SLR* checks its delay, the

[16]Postponing one query does not delay the others.

difference between its end time and the current system time, and sets the $t_{d_{max}}(p_n)$ to the maximum delay time of p_n events encountered so far. When smaller maximum delay times of events are assumed, queries are evaluated sooner and hence the overall system works in more real-time fashion, but there is more chance of answering a query when the complete history of events asked by the query is not in place yet. When larger maximum delay times of events are assumed, there is a higher chance to have all sensory data up to the time specified by the query already processed by the corresponding components and their results received by *SLR* when the query is evaluated. However, queries are performed with more delays.

The other way that *SLR* can ensure to have received the full history of events of a type p_n up to a time z in its knowledge base is by being told so by a component generating such events using special $updated(p_n)^z$ events. Whenever *SLR* receives such an event, it assumes that the history of events of the type p_n up to time z is available.

The query synchronization is often required for a query that interpolates the value of an attribute at a given time using *next* and *prev* term. The value can be interpolated as soon as the first relevant event after that time is received. *SLR* monitors memory events, discussed in Section 5.1.4, and evaluates the postponed queries as soon as necessary events are received.

Example 7. When the position of an object O in the world coordination frame at a time T is queried by the rule (9), the query can be answered as soon as both camera to base and base to world relative positions at time T can be evaluated by rules (7) and (8). The former can be evaluated (i.e. interpolated) as soon as *SLR* receives the first $tf(\text{'}cam\text{'}, \text{'}base\text{'}, P)$ event with a start time equal or later than T. The latter can be evaluated as soon as the *SLR* receives the first $tf(\text{'}base\text{'}, \text{'}world\text{'}, P)$ event with the start time equal or later than T, or when it can ensure that no $tf(\text{'}base\text{'}, \text{'}world\text{'}, P)$ event has occurred within $[T, T+1]$. If we assume $t_{d_{max}}(tf(\text{'}base\text{'}, \text{'}world\text{'}, P))$ is set to 0.5 second, *SLR* has to wait 1.5 second after T to ensure this.

Example 8. The robot is asked about the objects it sees on *table1*. To answer the question, the robot takes a number of pictures from the table starting at time t_1 and finishing by time t_2 and then the *SLR* knowledge base is queried by $\langle goal, t_2\rangle$ where the *goal* is

$$findall(obj(O, Type, L),\\ \quad (\ mem(obj(O), segR(O,L)^{T_x}),\ t_1 \leq T_x \leq t_2, prev(obj(O, Type, P)^{T_y}, t_2)\ ),\\ \quad List) \tag{12}$$

The query result is the list *List* of terms of the form *obj(O,Type,L)* matching the template specified by the second argument of the *findall* term. This includes all object segments recorded as $segR(O,L)^{T_x}$ events in $obj(O)$ memory instances recognized during $[t_1, t_2]$. The type of each object segment o_i is recognized by querying the last $obj(o_i,Type,P)$ event occurring before or at time t_2. To list all the objects, *SLR* makes sure to evaluate the query after the histories of both *segR(O,L)* and *obj(O,Type,L)* events up to time t_2 are available. A signaling mechanism to realize this is as follows. After finishing the processing of each image taken at a time t and outputting the recognized object segments, the *segRec* component sends out the event $updated(segR)^t$. The *IEC* receives these events sending object segments whose type is not known and the $updated(segR)^t$ events to the *objRec* component. We assume events of each type are communicated among the components in order. The *objRec* component receives some object segments recognized at a time t, processes them in the order it receives them and sends the recognized types back to the *IEC*. Whenever the *objRec* processes an $updated(segR)^t$ event, it realizes that it has finished processing of the object segments recognized up to time t and generates an $updated(obj)^t$ event. Receiving $updated(segR)^t$ and $updated(obj)^t$ events, *SLR* is notified when the histories of both types of events up to time t_2 are available and then evaluates the query.

5.2 Discussion

The use of memory in existing research includes collecting data from various sources and in time, mediating as a shared resource for component interaction (i.e. blackboard architectural pattern [77]), refining data by various processes, and integrating various reasoning capabilities to maintain and query the robot's knowledge of the environment for task execution, human interaction and learning [13, 80, 65, 37, 36, 71, 72, 50, 49, 52, 56]. A large set of on-demand information processing requirements have been discussed elsewhere [49, 80].

A main concern in supporting on-demand information processing is the choice of language for representing and storing data. The choice of language and its execution system largely determines the extent to which various on-demand information processing requirements along data, process, memory and access dimensions are supported, perhaps the most important ones being knowledge modelling and reasoning. The advantage of non logic-based data representations, for example, using programming data structures in *CAST* [36, 71] and *GSM* [56], is the flexibility and efficiency in the representation and manipulation of amodal data such as image data and probability distributions. However the expressiveness of queries for information maintained by such systems is limited. An interesting approach is the XML data representation by *IDA* [80, 65] supporting *Xpath* queries [15], for example, to retrieve data of objects recognized with confidence of more than a threshold. The data representation in non-logic based systems is usually tightly related to the data representation used in their underlying framework and does not support logical reasoning. In *Retalis*, binary data is represented as *String*. This requires encoding binary data to the *Prolog String* format when importing a *ROS* message to *Retalis* and decoding it when the data is sent back to *ROS*, which is time consuming. However, one can maintain the actual binary objects in *c++* and manipulate handlers to the objects in *Retalis*.

A recent survey of existing robotic *information management* systems [49] shows that most systems rely on logical formalisms, mainly including declarative languages such as the *OWL*[17] language [57] based on Description logics [10] and/or rule-based languages such as the *SWRL*[18] language [45] for rule-based reasoning in *OWL* and *Prolog*. In particular, *OWL* is a popular choice to define ontologies of various types of knowledge such as knowledge of space, objects, actions and robot capabilities used, for instance, in *ORO* [50, 49], *KnowRob* [72, 71] and *OUR-K* [52]. Defining ontologies are necessary to integrate various sources of knowledge such as the domain and common sense knowledge as performed by the aforementioned systems and for sharing robots' knowledge, for instance, in the cloud [73]. While we did not address modeling of knowledge, existing ontologies can be directly used in *Retalis* as *OWL* ontologies can be represented and reasoned upon in *Prolog*. For example, *KnowRob* offers one of the most comprehensive robotic ontologies and uses the *Prolog Semantic Web Library*[19] [60] for loading and storing *RDF*[20] [21] triples and the *Thea*[21] *OWL* parser library [75] for OWL reasoning on top of this representation.

The use of *Prolog* as the underlying technology for maintaining robotic *OWL* knowledge has a few practical advantages for inference compared to the use of existing description logic reasoners such as the *Pellet*[22] reasoner [68] used in *ORO*.

[17] http://www.w3.org/TR/2004/REC-owl-ref-20040210/
[18] http://www.w3.org/Submission/SWRL/
[19] http://www.swi-prolog.org/pldoc/package/semweb.html
[20] http://www.w3.org/RDF/
[21] http://www.semanticweb.gr/thea/
[22] http://clarkparsia.com/pellet/

Those reasoners keep a classified version of the knowledge base in memory specifying each individual belonging to which classes. Therefore continuous changes of the knowledge base through acquiring sensory data requires frequent re-classification of the whole knowledge which can be costly [72]. This problem can be partially addressed by optimizing this operation using an incremental updating technique [34]. The more important advantage is related to the open world assumption in *Description Logics* versus the closed world assumption in *Prolog*, and the monotonicity of description logics versus supporting a form of non-monotonicity in *Prolog* by the *negation as failure* inference rule within the closed world assumption. In the closed world assumption, representations can be more compact as 'a fact not being true' does not need to be described but it can be inferred by not being able to prove the fact. Moreover, the open world assumption and monotonicity of *Description Logic* makes the representation and reasoning on dynamics of the environment (i.e. changes and actions) difficult requiring to handle such aspects externally [84, 49], but, for instance, *KnowRob* implements a predicate to return an object's location at a time by searching for the last observation of the object's location before that time. Reasoning about changes and actions has been extensively studied in various knowledge formalisms such as Situation Calculus [51] and *Event Calculus* [47, 67]. The *SLR* language provides a practical and efficient solution for representing robot knowledge based on discrete observations, providing a means to deal with the temporal validity of data and representation of continuous domains which is not the focus of such formalisms. Compared to the *KnowRob* approach of, for instance, implementing a predicate to represent an object's location at a time, *SLR* simplifies the definition of such predicates in general and increases the efficiency of their computations by maintaining the sorted list of events based on their occurrence times. *Prolog* provides a flexible support for access to external data or reasoning functionalities while reasoning on knowledge through procedural attachments to the *Prolog* terms. This feature is used in *KnowRob*, for instance, to compute spatial relations between objects and in *Retalis* to integrate *OpenGL Mathematics*[23] (GLM) for arithmetic operations.

To the best of our knowledge the *SLR* support for synchronization of queries on knowledge built upon asynchronous data is not presented elsewhere. However, similar synchronization mechanisms as found in *SLR* are implemented in other robotic software in a more limited context. One example is the *DyKnow* framework [38] that synchronizes data received from streams of data based on different policies to generate new ones. Another example is the *tf* library [32] widely used in *ROS* for querying position transformation between robot's coordination frames over time.

[23]http://glm.g-truc.net/0.9.5/index.html

When a relative position at a time is queried, the query is not answered until receiving the first observation of that position at or after that time. The *tf* library only supports interpolation of data similar to the *SLR* rule (7). Therefore, even if a position is constant in time, its value needs to be continuously published to *ROS* consuming the network bandwidth. Moreover, sometimes a component such as *AMCL* in *ROS* provides updates in a slow rate but they are precise enough to be used until the next update is made available. In order to not delay the processing of data until availability of the next update, this component stamps its updates in the future.[24] Apart from being semantically confusing, time stamping updates in future can result in using old data even if new data is already available. With the *SLR* extrapolation approach, for instance, implemented by the rule (8), if a position transformation is static, its value does not need to be published being extrapolated from its last observed value. In addition, the time bound of the *next* predicate in *SLR* allows to specify how long *SLR* needs to wait to see whether a value has been changed, assuming after each relevant change a notification is received.

Except a few, most information management systems leave pruning data from the memory to external components. In *ORO*, knowledge is stored in different memory profiles, each keeping data for a certain period of time. In *IDA*, scripts are activated periodically or in response to events of memory changes to perform garbage collection. In *SLR*, flexible garbage collection functionalities are blended in the syntax of the language. In addition, a subtle difference between *SLR* and other systems is that in the existing systems, external components store the data in memory. In *SLR*, memory instances are declaratively defined which selectively store data from the input flow of events to the *SLR*. The storage of data in *SLR* is similar to active memories such as the ones of *IDA* and *CAST* as data is recorded in memory instances with unique identifiers, however *SLR* supports logical reasoning over the contents of memory instances. This approach supports having different memory profiles for different pieces of data and a flexible way of selecting the data that are to be reasoned about as a whole, thus allowing to reason about a part of knowledge that could be inconsistent with other part of the knowledge maintained in the memory. Furthermore, active memories allow external components to update the contents of memory instances. As such, suitable error handling and locking mechanisms are necessary to synchronize the parallel access to memory. In contrast, the modeling of the input as a stream of events and clear semantics of memory instances in *SLR* removes much of the problems related to the parallel access of data. For an example, consider two components processing object segment events to recognize the orientation and type of objects. In our approach, this can be

[24] http://wiki.ros.org/tf/FAQ

implemented as follows: an object segment event is sent to both components, these components perform their processes and generate their uniquely typed events. Then an *ELE* rule receives events from these components, synchronizes them based on their object identifiers and occurrence times and produces new events of recognized objects with their types and orientations. In a naive approach, object segments are recorded in the memory and are processed and updated by both components in parallel which could re-write each other results.

SLR supports notifying external components when memory instances are added or deleted to the memory. This can be easily extended to also generate corresponding notifications when events are added or deleted from memory instances. However, the input flow of events to *SLR* is processed by *ELE*. Therefore external components can subscribe to *Retalis* to be notified when the data of interest is being fed to *SLR*. While notifying changes of the memory is a main functionality in active memories, it is less common in logic-based knowledge management systems. An exception is *ORO* to which one can subscribe to receive a notification, whenever a fact can be inferred by the *ORO* knowledge base. However it is not described whether or not this includes the knowledge that can be derived by *SWRL* rules. Moreover, it not described whether this functionality is implemented by continuously querying the knowledge base for such a fact, or it is efficiently realized by an incremental and event-driven algorithm such as backward chaining rules in *ELE* [6, 5, 3].

6 Evaluation

This section evaluates the performance of *Retalis* by demonstrating the implementation of an application for a NAO robot. NAO is a small programmable humanoid robot offered by Aldebaran Robotics[25], equipped with advanced sensors such as cameras, touch sensors and microphones. In the application, NAO observes objects in the environment, perceiving their relative positions to its camera, and computes the position of objects in the environment. Figure 5 presents software components[26] of the NAO application, operating as follows. The *NAO nodes*[27] component provides an interface to acquire sensory data and to command the NAO robot. It publishes images generated by the top camera of the robot. It also publishes events about the transformation among the robot's coordinate frames. Each of these events contains a set of transformations where each transformation specifies the relative position

[25] http://www.aldebaran.com/en

[26] The software includes also a face recognition component which is not discussed for brevity.

[27] http://wiki.ros.org/nao_robot

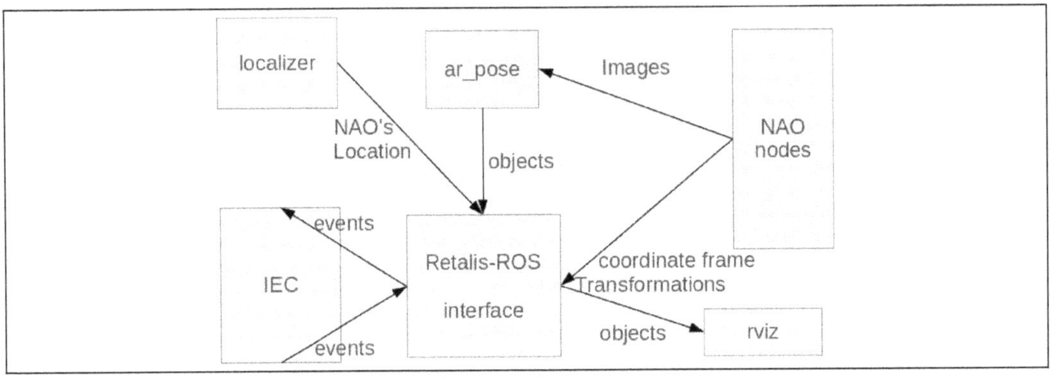

Figure 5: NAO's software components

among two coordinate frames. The *ar_pose*[28] component processes the images to recognize objects and calculates the position of objects with respect to the camera. Each event from *ar_pose* contains data of a set of observed objects. The *localizer* component calculates the robot's position in the world. The *IEC* component is subscribed to information about objects' positions, robot's location and coordinate transformations. It calculates the position of objects in the world from the transformation among the following pairs of coordinate frames, *(world, base_link), (base_link, torso), (torso, neck), (neck, camera) and (camera, object).* The arithmetic operations are performed using the *OpenGL Mathematics*[29] (GLM) library which has been integrated in *Retalis.* The *rviz*[30] component visualizes the objects in the environment. The *IEC* communication with other nodes is realized by the *Retalis-ROS interface* component. This component converts *ROS* messages to *Retalis* events and vice versa. The *IEC* and the *Retalis-ROS interface* components are implemented in *Retalis.*

6.1 Basic setup

For a first test implementation, all software components run remotely on an XPS Intel Core i7 CPU@ 2.1 GHz x 4 laptop running ubuntu 12.04 LTS, connected to the NAO robot. After the evaluation phase, the software will be implemented in the NAO robot itself. NAO comes with an Intel Atom CPU@1.6 GHz running Linux. The performance is evaluated by measuring the CPU time, the amount of time of a CPU of the computer that is used by the *Retalis* program. We measure the CPU time

[28] http://wiki.ros.org/ar_pose
[29] http://glm.g-truc.net/0.9.5/index.html
[30] http://wiki.ros.org/rviz

as the percentage of the CPU's capacity (i.e. CPU usage percentage) computed by the operating system. In the following graphs, the vertical axis represents the CPU usage percentage and the horizontal axis represents the running time in seconds. The CPU time is logged every second and is plotted using "gnuplot smooth bezier".

The NAO application includes the following tasks:

- On-flow processing: events from *ar_pose* and *NAO nodes* are split into respective events such that each event contains data of a single object or the transformation among a single pair of coordinate frames. The transformation data among pairs of coordinate frames are published with frequencies from 8 to 50 hertz. There are in average 7 objects perceived per second. In total, *Retalis* processes about 1900 events per second.

- Memorizing and forgetting: there are 5 memory instances observing the events. They record and maintain the last 30 seconds histories of the transformation among the pairs of coordination frames used to calculate the transformation among *world* and *camera*.

- Querying memory instances: for each observed object, *SLR* is queried for the *world-to-camera* transformation. The transformation among a pair of coordinate frames at a time is calculated by interpolation, as performed by the rule (11). Each interpolation requires accessing a memory instance twice, once using a *prev* term and once using a *next* term. To calculate the position of all objects, memory instances are accessed 70 times per second.

- Synchronization: a query is delayed in case any of the necessary transformations can not be interpolated from the data received so far. *Retalis* monitors the incoming events and performs the delayed queries as soon as all data necessary for their evaluations are available.

- Subscription: there are 8 distinct objects in the environment and consequently 8 subscriptions to publish recognized objects to distinct *ROS* topics. The *rviz* component is subscribed to these topics to visualize the position of objects.

Figure 6 shows the CPU time used by the *Retalis* and *Retalis-ROS*-converter nodes when running the NAO application. The *Retalis* node calculates the position of objects in real-time. It processes about 1900 events, memorizes 130 new events and prunes 130 outdated events per second. It also queries memory instances, 70 times per second. These tasks are performed using about 18 percent of the CPU time. In this experience, the *Retalis* node has been directly subscribed to *ROS* messages containing information about coordinate transformations and recognized

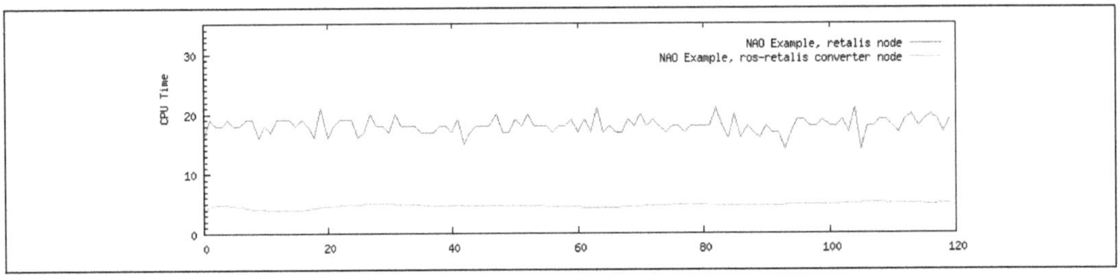

Figure 6: NAO application

objects. The *Retalis-ROS*-converter, consuming about 5 percent of CPU time, only subscribes *Retalis* to the recognized faces and converts and publishes events about objects' positions to *ROS* topics.

As we saw in Section 3, *Retalis* provides an easy way to subscribe to *ROS* topics and automatically convert *ROS* messages to events. This is implemented by the *Retalis-ROS*-converter node. The implementation is in Python and is realized by inspecting classes and objects at runtime and therefore is expensive. Figure 7 shows the CPU time used by the *Retalis* and *Retalis-ROS*-converter nodes for the NAO application, when the *Retalis-ROS*-converter is used to convert all *ROS* messages to *Retalis* events. In the previous configuration, the conversion from *ROS* messages, containing information about coordinate transformations and recognized objects, to events was performed by a manually written *c++* code, rather than using the *Retalis* automatic conversion functionality written in *Python*. We observe that in the new configuration, the *Retalis* node consumes a few percent less, but the *Retalis-ROS*-converter node consumes about forty percent more CPU time, comparing to the previous configuration. These results show that while the automatic conversion among messages and events are desirable in a prototyping phase, the final application should implement it in C++ for performance reasons. We will investigate the possibility to optimize and re-implement the *Retalis-ROS*-converter node in C++.

Metric evaluation of languages and systems like *Retalis*, in general, is challenging for the following reasons[49, 48, 56]. Experiments often involve many other modules running in parallel and building repeatable experiments for robots in dynamic environments is challenging. In addition, very few existing systems report metric evaluations and the lack of standard *API*s and differences in functionalities makes it hard to compare these systems. The rest of this section evaluates main *Retalis* functionalities. We report a number of experiments using data from the NAO application, recorded by *rosbag*.[31] Using *rosbag*, data can be played in a simulation, as if it is played in real-time. While single performance results in the following ex-

[31] http://wiki.ros.org/rosbag

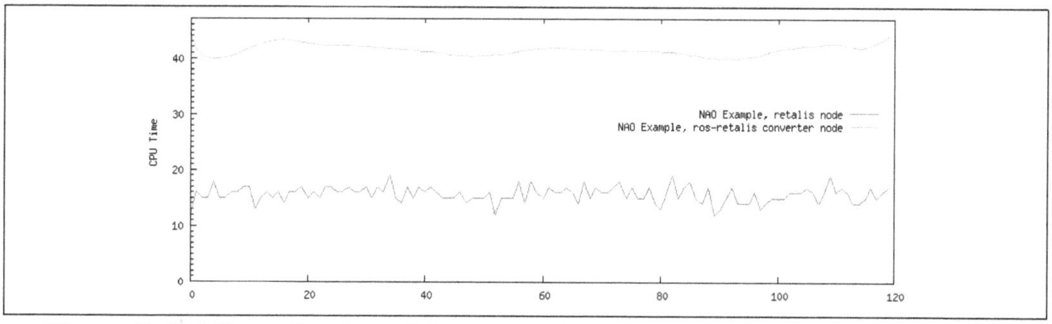

Figure 7: NAO application with automatic conversion of messages and events

periments depend on the NAO application, a series of experiments is presented for each functionality, allowing us to make a number of general observations about the performance of *Retalis* functionalities.

6.2 Forgetting and Memorizing

This section evaluates the performance of the memorizing and forgetting functionalities. We measure the CPU time for various runs of the NAO application where the numbers and types of memory instances are varied. We discuss the performance of memory instances by comparing the CPU time usages in different runs.

When an event is processed, updating memory instances includes the following costs:

- Unification: finding which memory instances match the event.
- Assertion: asserting the event in the database for each matched memory instance.
- Retraction: retracting old events from memory instances that reached their size limit.

Figure 8 shows the CPU time for a number of runs where up to 160 memory instances are added to the NAO application. These memory instances record *a(X, Y, Z, W)* events. Among the events processed by *Retalis*, there are no such events. The results show that the increase in CPU time is negligible. This shows that a memory instance consumes CPU time only if the input stream of events contains events whose type matches the type of events the memory instance records.

In Figure 9, the green and blue lines show the CPU time for cases where 20 memory instances of type tf(X,Y,V,Q) are added to the NAO application. These

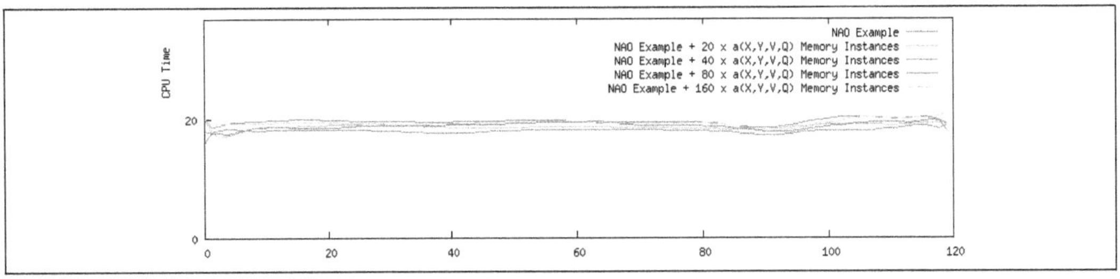

Figure 8: Irrelevant memory instances

memory instances match all tf events, about 1900 of such is processed every second. The size of memory instances for the green line is 2500. These memory instances reach their size limit in two seconds. After this time, the CPU time usage is constant over time and includes the costs of unification, assertion and retraction for updating 20 memory instances with 1900 events per second. The size of memory instances for the blue line is 150,000. It takes about 80 seconds for this memory instances to reach their size limit. Consequently, the CPU time before the time 80 only includes the costs of unification and assertion, but not the costs of retraction. After the time 100, the CPU usages of both runs are equal. This shows that the cost of a memory instance does not depend on its size.

The purple line shows the CPU time for the case where similarly there are 20 memory instances of type tf(X,Y,V,Q). However, these memory instances record events until they reach their size limit. We added a condition for these memory instances such that after reaching their size limit, they perform no operation when receiving new events. After the time 100, the CPU time is constant about 23 percent, being 5 percent more than the CPU time of the NAO application, represented by the red line. This 5 percent increase represents the unification cost. This also shows that the costs of about 38000 assertions and 38000 retractions per second is about 30 percent of CPU time. In other words, 2500 memory updates (i.e. assertions or retractions) are processed using one percent of CPU time.

Figure 10 shows the CPU time for a number of runs where up to 40 memory instances of type tf(X,Y,Z,W) and size 2500 are added to the NAO application. The red line at the bottom shows the CPU time for the NAO application. We make the following observations. Adding first 10 memory instances to the NAO application increases the CPU time about 20 percent. After that, adding each set of 10 memory instances increases the CPU time about 13 percents. This shows that the cost grows less than linearly. The implementation of memory instances is in a way that the cost of an assertion or a retraction can be assumed constant. This means that the unification cost for the first set of memory instances is the highest. In other words,

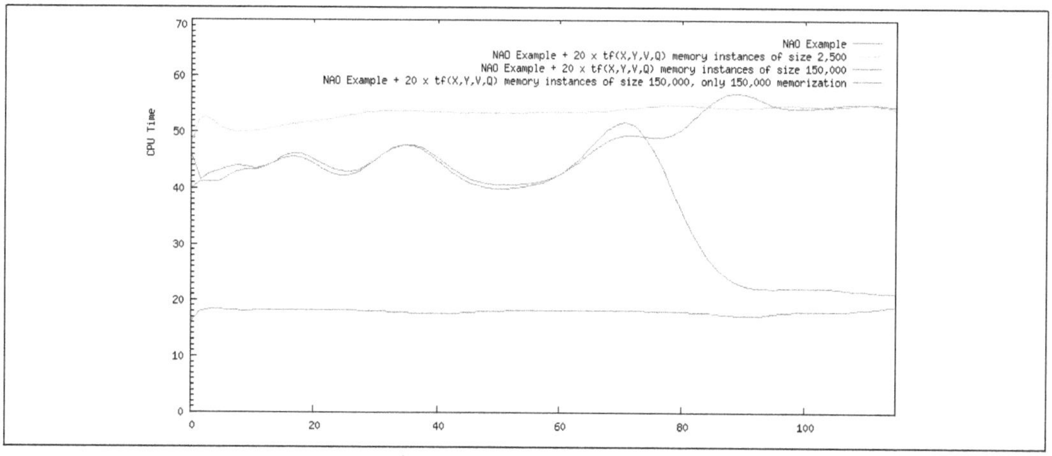

Figure 9: tf(X,Y,V,Q) memory instances (1)

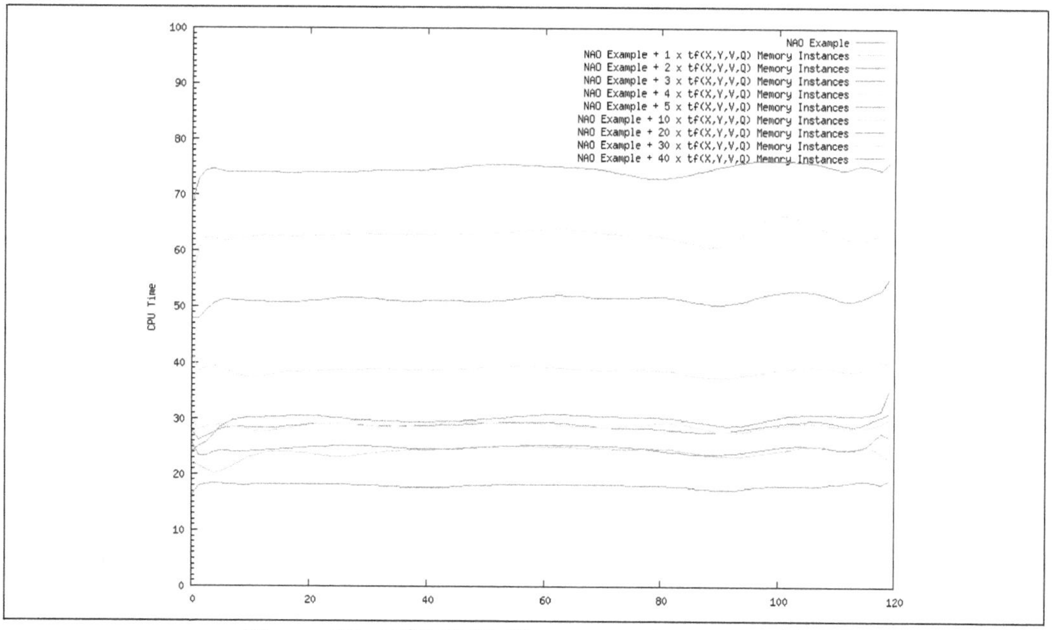

Figure 10: tf(X,Y,V,Q) memory instances (2)

the unification cost per memory instance decreases when the number of memory instances are increased. The reason relates to the way that the underlying *SWI-Prolog* engine searches and unifies terms which is not investigated here.

Figure 11 shows the CPU time for a number of runs where up to 640 memory instances of type tf(head,camera,Z,W) and size 2500 are added to the NAO applica-

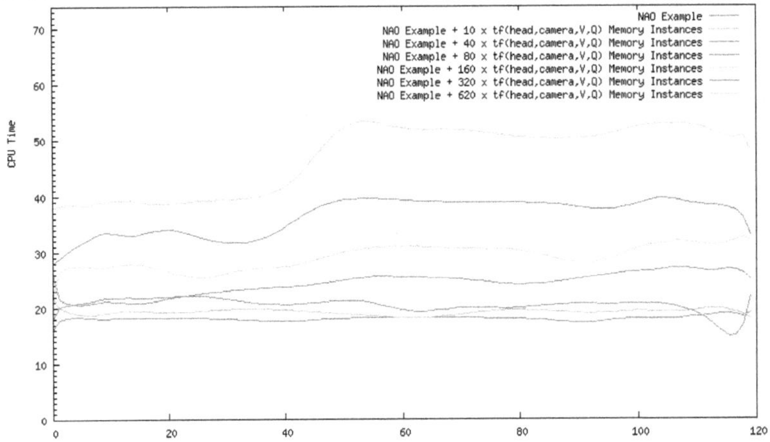

Figure 11: tf(head,cam,V,Q) memory instances

tion. The events matching these memory instances are received with the frequency of 50 Hz. We make the following observations. First, it takes 50 seconds for these memory instances to reach their size limit. After 50 seconds, these memory instances reach their maximum CPU usages, as the costs of retraction is added. Second, each memory instance filters 1900 events per second recording about two percent of them. The cost of 640 memory instances is about 35 percent of CPU time. Third, the unification cost per memory instance is decreased when the number of memory instances are increased.

Figure 12 compares the costs of different types of memory instances. The purple line shows the CPU time for the case where there are 10 memory instances of type tf(X,Y,V,Q). The green line shows the CPU time for the case where there are 320 memory instances of type tf(head,cam,V,Q). We observe that the costs of both cases are equal. The memory instances in the former case record 19,000 events per second (i.e. 10*1900). The memory instances in the latter case filter 1900 events per seconds for tf(head,cam,V,Q) events, recording 16000 events per second (i.e. 320*50). The results show the efficiency of the filtering mechanism.

The brown line shows the CPU time for the case where there are 10 memory instances of type tf(X,Y,V,Q) and 320 memory instances of type tf(head,cam,V,Q). Comparing it with the green and purple lines shows that the CPU time usage of these memory instances is less than sum of the CPU usages by 10 tf(X,Y,V,Q) memory instances and 320 tf(head,cam,V,Q) memory instances. This shows that the unification cost per memory instance is decreased when the number of memory instances are increased, even when the memory instances are not of the same type.

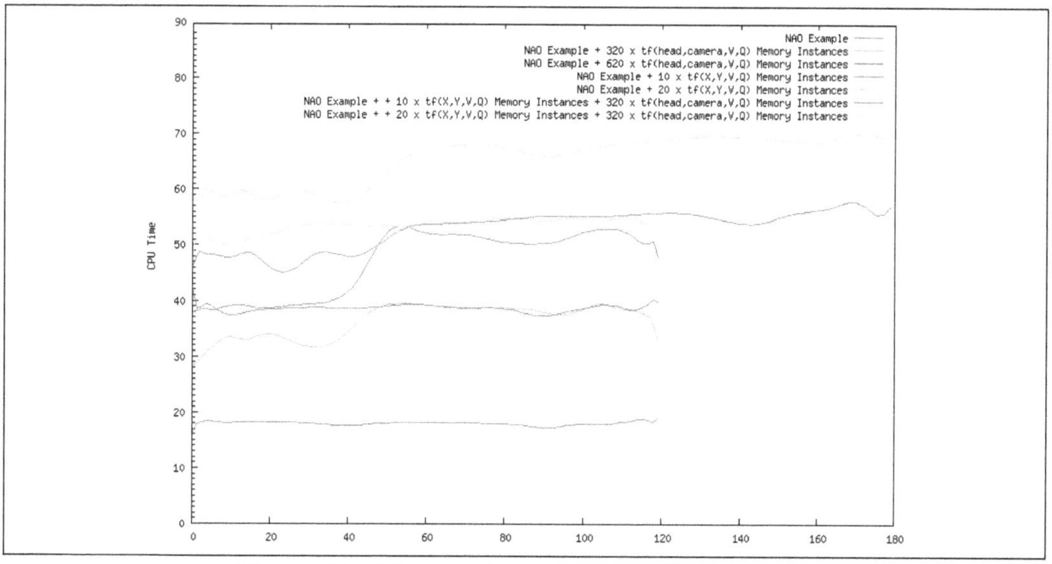

Figure 12: Memory instances of different types

These experiments show that *Retalis* is able to maintain a history of a large volume of data. Memorizing and forgetting functionalities of *SLR* have been optimized as follows. A memory instance memorizes an event by creating an event record containing the event and the identifier of the memory instance. The event record is asserted as the top fact in the database. This operation takes a constant time. Event records of a memory instance are numbered in order of the event occurrence times. *SLR* generates a hash key for each event record, based on the respective identifier and the record number. Event records are indexed on their hash keys. Consequently, accessing an event record takes a constant time *SLR* keeps track of the number of the oldest event record of each memory instance. Therefore, forgetting takes a constant time, irrelevant of the size of memory instances.

6.3 Querying

Retalis queries are *Prolog*-like queries executed by the *SWI-Prolog* system. The following evaluates the performance of next and prev terms and the synchronization mechanism which are specific to *Retalis*. The performance of next and prev terms are important because the sensory data recorded by *Retalis* is queries using these terms. Not only does *Retalis* extend the *Prolog* language with these built-in terms to provide easier syntax for querying history of data, but also to make querying of data more efficient.

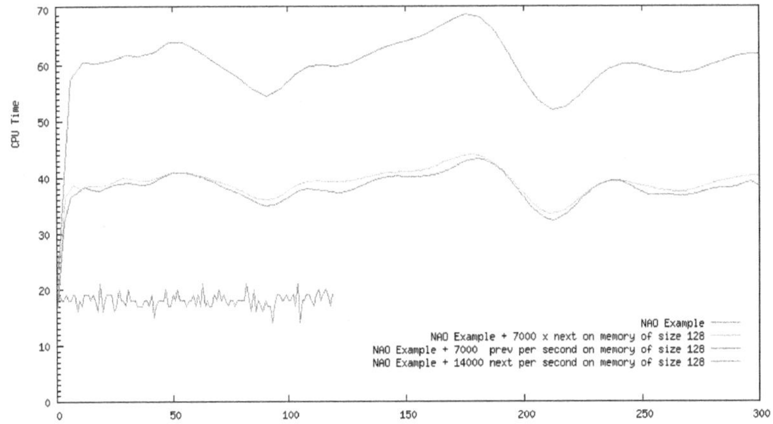

Figure 13: Next and prev terms (1)

Querying Memory Instances

This section evaluates the performance of prev and next terms used to access event records in memory instances. *Retalis* optimizes the evaluation of these terms as follows. It keeps track of the number of event records in each memory instance. The prev and next terms are evaluated by a binary search on event records. An access to an event record by its number takes a constant time. Consequently, the evaluation of prev and next is done in logarithmic time on the size of the respective memory instance. In Figures 13, 14 and 15 below, the red line visualizes the CPU time of the NAO application.

The green line in Figure 13 visualizes the CPU time of the NAO application adapted as follows. There is an additional tf(head,cam,V,Q) memory instance of size 128. This memory instance is queried by 1000 next terms for each recognition of an object. In average, 7000 next terms are evaluated per second. The blue line visualize the CPU time of a similar program in which 7000 prev terms are evaluated per seconds. The figure shows that the costs of the evaluations of prev and next terms are similar. The purple line shows the CPU time of the case where 14,000 next terms are evaluated per second. We observe that the cost grows linearly.

The blue line in Figure 14 visualizes the CPU time of the case where 7000 next terms are evaluated per second. The green line visualizes the CPU time of the case where there are 320 tf(head,cam,V,Q) memory instances added to the NAO application. The purple line visualizes the CPU time of the case where 7000 next terms are evaluated per second and there are 320 tf(head,cam,V,Q) memory instances. We observe that the cost of accessing a memory instance does not depend on existence

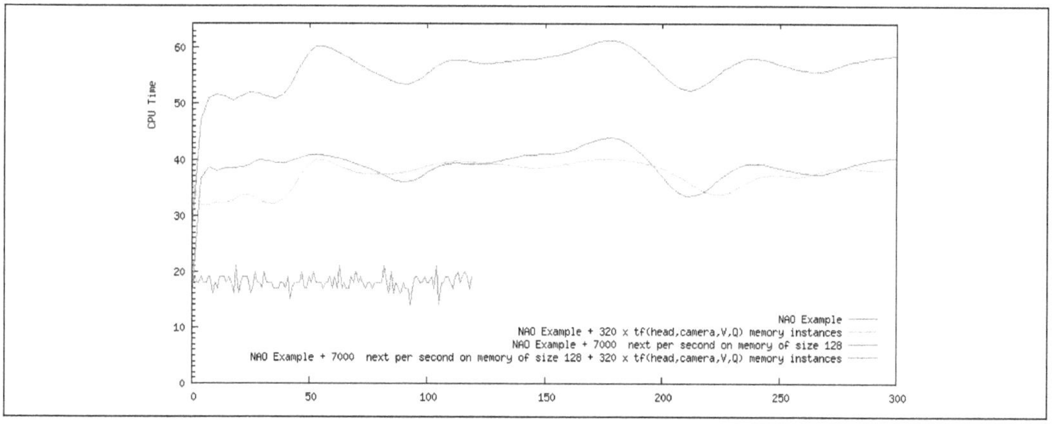

Figure 14: Next and prev terms (2)

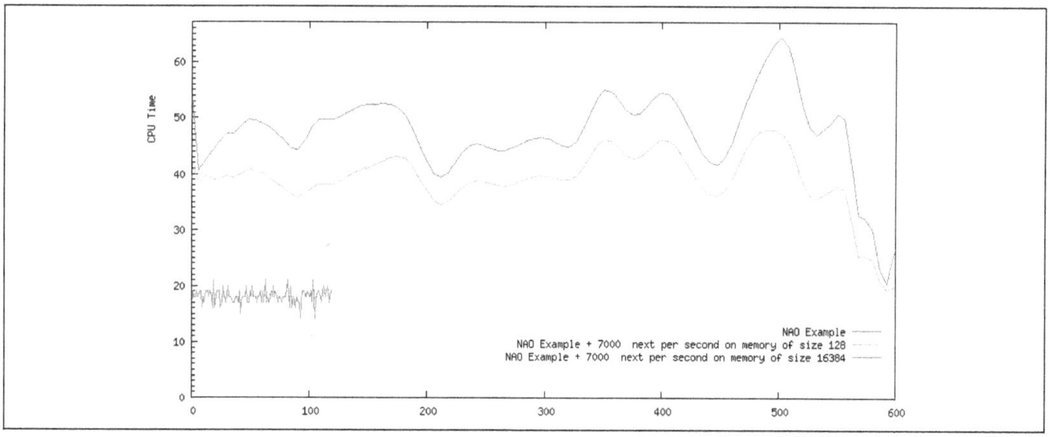

Figure 15: Next and prev terms (3)

of other memory instances.

The green line in Figure 15 visualizes the CPU time of evaluating 7000 next terms per second on a memory instance of size 128. The blue linevisualizes the CPU time of evaluating 7000 next terms per second on a memory instance of size 16384. The size of the memory instance in the latter case is the power of two of the size of the memory instance in the former case. The increase in the CPU time for the latter case, with respect to the NAO application, is less than two times of the increase in the CPU time for the former case.

The *prev* and *next* terms provide efficient ways of accessing records of events. Otherwise, all event records should be read, for instance, to find the latest position of an object. For example, an experiment is reported for the *KnowRob* knowledge

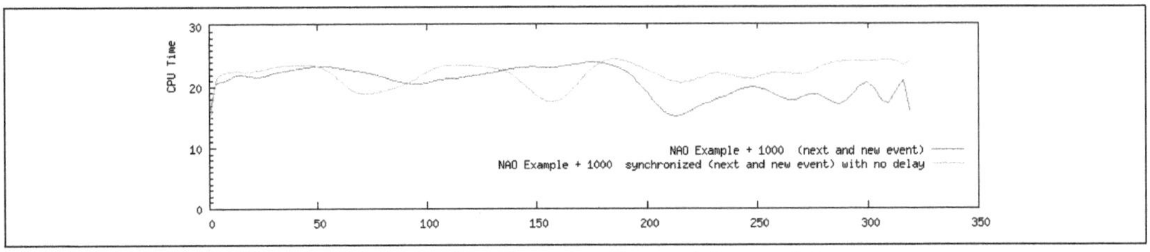

Figure 16: Synchronization with no delay

base where there are 65,000 records of events about the location of an object. It takes 11 seconds to find the latest location [72].

Synchronization

The synchronization mechanism is implemented as follows. Before evaluating a query, memory instances are checked whether they are up-to-date with respect to the query (i.e. the query is achievable as defined in Section 5.1.5). If the query cannot be evaluated, it is recorded as a postponed query. For each postponed query, *Retalis* generates a set of monitors. Monitors observe memory update events. As soon as all necessary events are in place in memory instances, the query is performed. The implementation of monitors are similar to the implementation of memory instances.

The red line in Figure 16 visualizes the CPU time of the NAO application where in each second, 1000 next queries on a memory instance of size 2500 are evaluated. In addition, for each next query, a new event is generated. The green line visualizes the CPU time of a similar case where the next queries are synchronized. This experiment is conducted in a way that no query needs to be delayed. Comparing these two cases shows that when queries are not delayed, the synchronization cost is negligible.

Figure 17 shows the CPU time of four cases. In all these cases, 1000 synchronized next queries are evaluated and 1000 events are generated in each second. The red line visualizes the case where no query is delayed. The green line visualizes the case where queries are delayed for 5 seconds. In this case, the memory instance queried by a next term has not yet received the data necessary to evaluate the query. The query is performed as soon as the memory instance is updated with relevant information. There are 1000 queries per seconds, each delayed for 5 seconds. This means there exist 5000 monitors at each time. These monitors observe 1900 events processed by *Retalis* per second. We observe that for such a large number of monitors observing such a high-frequency input stream of events, the increase in CPU time is less than 30 percent.

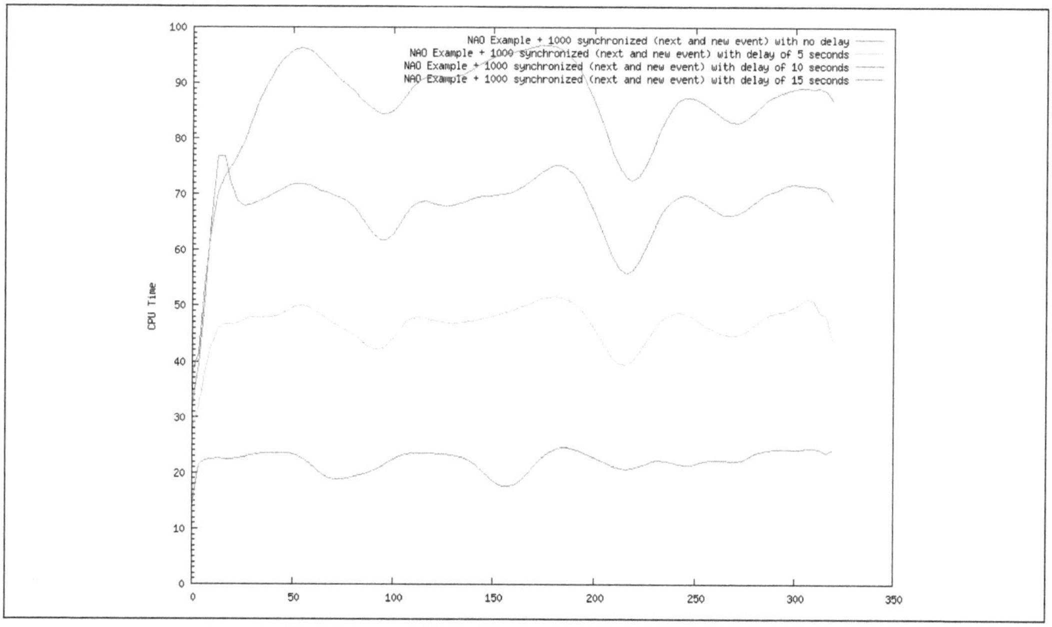

Figure 17: Synchronization with delays

6.4 On-Flow Processing

On-flow processing functionalities in *Retalis* are implemented using *ELE*. *ELE* execution model is based on decomposition of complex event patterns into intermediate binary event patterns (goals) and the compilation of goals into *goal-directed event-driven Prolog* rules. As relevant events occur, these rules are executed deriving corresponding goals progressing toward detecting complex event patterns.

Information flow processing systems such as *ELE* are designed for applications that require a real-time processing of a large volume of data flow. We refer the reader to the evaluation of the performance of *ELE* presented elsewhere [6, 3]. While the execution system of *ELE* is *Prolog*, the evaluation shows that in terms of performance, *ELE* is competitive with respect to the state-of-the-art information processing systems.

6.5 Subscription

The implementation of the subscriptions is similar to the implementation of memory instances. The only difference is that an event matching a memory instance is asserted to the knowledge base for that memory instance, and an old event is retracted if the memory instance is full, but an event matching a subscription is delivered to the respective subscriber. Consequently, the costs of subscriptions include the unification cost, discussed in section 6.2, and the costs to publish events to subscribed *ROS* topics. The latter comprises the costs for converting events to *ROS* messages and the costs of message transportation within the *ROS* framework.

7 Conclusion

Retalis is introduced in this paper to develop information engineering components of autonomous robots. Consequently it is used for the processing and management of data to create knowledge about the robot's environment. Information engineering is an essential robotic technique to apply AI methods such as situation awareness, task-level planning and knowledge-intensive task execution. Consequently, *Retalis* addresses a major challenge to make robotic systems more responsive to real-world situations.

The *Retalis* language integrates *ELE* and *SLR*, two logic-based languages for on-demand and on-flow processing, respectively. *ELE* is used for temporal and logical reasoning, and data transformation in flow of data. *SLR* is used to implement a knowledge base maintaining history of some events. *SLR* supports state-based representation of knowledge built upon discrete sensory data, management of sensory data in active memories and synchronization of queries over asynchronous sensory data.

Retalis addresses all eight requirements discussed in the introduction. In particular, *ELE* addresses the requirements of on-flow processing, like event-driven and incremental processing, temporal pattern detection and transformation, subscription and garbage collection. *SLR* addresses the requirements of on-demand processing like memorizing, forgetting, active memory and state-based representation. In this way, *Retalis* unifies and advances the state-of-the-art research on robotic information engineering.

The contribution of this paper is threefold. The first contribution is the development of *SLR* language. *SLR* advances the state-of-the-art robotic on-demand processing systems by providing an active logic-based knowledge base. It combines the benefits of both knowledge base and active memory systems. *SLR* provides programming constructs to facilitate a high-level and efficient implementation of robotic on-demand processing functionalities. However, *SLR*, is a logic-based language based on *Prolog*. Therefore, a knowledge of *Prolog* is necessary to use *SLR*.

The second contribution is the integration of the *ELE* and *SLR* languages concerning three issues. The first issue is to process flows of sensory data on the fly by *ELE* to extract relevant knowledge for its compact storage in *SLR*. The second issue is to query *SLR* for the knowledge built upon sensory data while processing flows of data. The third issue is to process events of changes of *SLR* memory by *ELE* to notify external components with patterns of changes that are of their interest.

The third contribution of the declarative *Retalis* language is a semantics based on a model of sensory data taking into account their occurrence times. This may be contrasted to alternative semantics based on processing times. In this way, the model captures and handles various issues related to asynchronous processing of data in robot software.

Moreover, *Retalis* is an open-source and framework-independent software library. Therefore, it can be used to empower the existing robotic frameworks with its wide range of functionalities as opposed to, for instance, robotic active memories which are usually tightly integrated with specific robotic frameworks. *Retalis* has been integrated in *ROS* and used to implement few proof-of-concept tasks for NAO robot, including data transformation, runtime subscription and high-level event detection.

A future work is to apply the machinery developed in this paper to support AI-based robotics. For example, AI research on task-level planning has developed a number of agent programming languages [17] to support the implementation of autonomous behavior based on the BDI (Belief-Desire-Intention) model of practical reasoning [18, 64, 63]. However, these languages do not support event-driven and incremental reasoning on their input data. Therefore, the sensory input processing support of these languages is not suitable for on-flow processing of data [82]. The lack of on-flow processing support reduces the reactivity and limits the application of these languages in robotics [82, 81]. Moreover, there are concerns about the performance of these languages in robotic applications. For instance, there is a performance issue caused by the repetition of queries on knowledge base. An approach to increase performance is to cache query results [1]. By caching, a query is re-evaluated only if the knowledge base has been updated with relevant facts. To implement such a caching mechanism when the agent and the knowledge base components are separated, active memory functionalities are required to inform the agent program about the changes of the knowledge base.

Other future work is to further support on-flow and on-demand processing. A work is to support the representation and reasoning about uncertain data. Very few current information engineering systems address uncertainty. This can be due to scalability issues and the training time required to learn the transition probabilities of a domain. Another work is to extend consumption, and memorizing and forgetting policies in on-flow and on-demand processing, respectively. Another work is to further support reasoning on temporally qualified knowledge. Supporting the implementation of episodic-like memories [70] is another direction of research.

References

[1] Natasha Alechina, Tristan Behrens, Mehdi Dastani, Koen Hindriks, Koen Hubner, Fred Jomi, Brian Logan, Hai H. Nguyen, and Marc van Zee. Multi-cycle query caching in agent programming. In *Twenty-Seventh AAAI Conference on Artificial Intelligence (AAAI-13)*, July 2013.

[2] James F. Allen. Maintaining knowledge about temporal intervals. *Communications of the ACM*, 26(11):832–843, November 1983.

[3] Darko Anicic. Event Processing and Stream Reasoning with ETALIS. *PhD Thesis, Karlsruher Institute of Technology*, 2011.

[4] Darko Anicic, Paul Fodor, Sebastian Rudolph, and Nenad Stojanovic. Ep-sparql: A unified language for event processing and stream reasoning. In *Proceedings of the 20th International Conference on World Wide Web*, WWW '11, pages 635–644, New York, NY, USA, 2011. ACM.

[5] Darko Anicic, Paul Fodor, Sebastian Rudolph, Roland Stühmer, Nenad Stojanovic, and Rudi Studer. A Rule-Based Language for Complex Event Processing and Reasoning. In Pascal Hitzler and Thomas Lukasiewicz, editors, *Web Reasoning and Rule Systems SE - 5*, volume 6333 of *Lecture Notes in Computer Science*, pages 42–57. Springer Berlin Heidelberg, 2010.

[6] Darko Anicic, Sebastian Rudolph, Paul Fodor, and Nenad Stojanovic. Real-time complex event recognition and reasoning - a logic programming approach. *Applied Artificial Intelligence*, 26(1-2):6–57, 2012.

[7] Krzysztof R Apt and M H van Emden. Contributions to the Theory of Logic Programming. *J. ACM*, 29(3):841–862, July 1982.

[8] Alexander Artikis, Georgios Paliouras, François Portet, and Anastasios Skarlatidis. Logic-based representation, reasoning and machine learning for event recognition. In *Proceedings of the Fourth ACM International Conference on Distributed Event-Based Systems - DEBS '10*, page 282, New York, New York, USA, 2010. ACM Press.

[9] Carlos Astua, Ramon Barber, Jonathan Crespo, and Alberto Jardon. Object detection techniques applied on mobile robot semantic navigation. *Sensors*, 14(4):6734–6757, 2014.

[10] Franz Baader, Ian Horrocks, and Ulrike Sattler. *Handbook of Knowledge Representation*, volume 3 of *Foundations of Artificial Intelligence*. Elsevier, 2008.

[11] Yaakov Bar-Shalom and Thomas E. Fortmann. *Tracking and Data Association*. Academic Press Professional, Inc, 1988.

[12] Davide Francesco Barbieri, Daniele Braga, Stefano Ceri, Emanuele Della Valle, and Michael Grossniklaus. Querying rdf streams with c-sparql. *SIGMOD Rec.*, 39(1):20–26, September 2010.

[13] C. Bauckhage, S. Wachsmuth, M. Hanheide, S. Wrede, G. Sagerer, G. Heidemann, and H. Ritter. The visual active memory perspective on integrated recognition systems. *Image and Vision Computing*, 26(1):5–14, January 2008.

[14] Michael Beetz, Lorenz Mosenlechner, and Moritz Tenorth. CRAM - A Cognitive

Robot Abstract Machine for everyday manipulation in human environments. In *2010 IEEE/RSJ International Conference on Intelligent Robots and Systems*, pages 1012–1017. IEEE, October 2010.

[15] Mark Birbeck. *Professional XML*. Wrox Press, 2001.

[16] Nico Blodow, Dominik Jain, Zoltan-Csaba Marton, and Michael Beetz. Perception and probabilistic anchoring for dynamic world state logging. *2010 10th IEEE-RAS International Conference on Humanoid Robots*, pages 160–166, December 2010.

[17] Rafael H Bordini, Lars Braubach, Jorge J Gomez-sanz, Gregory O Hare, Alexander Pokahr, and Alessandro Ricci. A survey of programming languages and platforms for multi-agent systems. *INFORMATICA*, 30:33–44, 2006.

[18] Michael E Bratman. *Intention, Plans, and Practical Reason*. Cambridge University Press, March 1999.

[19] Davide Brugali and Patrizia Scandurra. Component-based Robotic Engineering Part I : Reusable building blocks. *IEEE ROBOTICS AND AUTOMATION MAGAZINE*, XX(4):1–12, 2009.

[20] Davide Brugali and Azamat Shakhimardanov. Component-based Robotic Engineering Part II : Systems and Models. *IEEE ROBOTICS AND AUTOMATION MAGAZINE*, XX(1):1–12, 2010.

[21] K. Selçuk Candan, Huan Liu, and Reshma Suvarna. Resource description framework: Metadata and its applications. *SIGKDD Explor. Newsl.*, 3(1):6–19, July 2001.

[22] L. Chittaro and A. Montanari. Efficient temporal reasoning in the cached event calculus. *Computational Intelligence*, 12(3):359–382, August 1996.

[23] W. F. Clocksin and C. S. Mellish. *Programming in Prolog*. Berlin-New York: Springer-Verlag, 2003.

[24] Silvia Coradeschi and Alessandro Saffiotti. An introduction to the anchoring problem. *Robotics and Autonomous Systems*, 43(2-3):85–96, May 2003.

[25] Claudia Cruz, Luis Enrique Sucar, and Eduardo F Morales. Real-time face recognition for human-robot interaction. In *Automatic Face & Gesture Recognition, 2008. FG'08. 8th IEEE International Conference on*, pages 1–6. IEEE, 2008.

[26] Gianpaolo Cugola and Alessandro Margara. Processing flows of information: From data stream to complex event processing. *ACM Computing Surveys (CSUR)*, V(i):1–70, 2012.

[27] Daniel de Leng and Fredrik Heintz. Towards on-demand semantic event processing for stream reasoning. In *Information Fusion (FUSION), 2014 17th International Conference on*, pages 1–8. IEEE, 2014.

[28] Patrick Doherty, Fredrik Heintz, and Jonas Kvarnström. Robotics, Temporal Logic and Stream Reasoning. *International Conference on Logic for Programming, Artificial Intelligence and Reasoning (LPAR-19)*, pages 42–51, 2014.

[29] Patrick Doherty, Jonas Kvarnström, and Fredrik Heintz. A temporal logic-based planning and execution monitoring framework for unmanned aircraft systems. *Autonomous Agents and Multi-Agent Systems*, 19(3):332–377, February 2009.

[30] J. Elfring, S. van den Dries, M.J.G. van de Molengraft, and M. Steinbuch. Semantic world modeling using probabilistic multiple hypothesis anchoring. *Robotics and Autonomous Systems*, 61(2):95–105, December 2012.

[31] Patrick Th. Eugster, Pascal A. Felber, Rachid Guerraoui, and Anne-Marie Kermarrec. The many faces of publish/subscribe. *ACM Computing Surveys*, 35(2):114–131, June 2003.

[32] Tully Foote. tf: The transform library. In *Technologies for Practical Robot Applications (TePRA), 2013 IEEE International Conference on*, Open-Source Software workshop, pages 1–6, April 2013.

[33] Malik Ghallab. On Chronicles: Representation, On-line Recognition and Learning. In *Proceedings of the Fifth International Conference on Principles of Knowledge Representation and Reasoning (KR'96)*, pages 597–606, 1996.

[34] Christian Halashek-Wiener, Bijan Parsia, and Evren Sirin. Description Logic Reasoning with Syntactic Updates. In Robert Meersman and Zahir Tari, editors, *On the Move to Meaningful Internet Systems 2006: CoopIS, DOA, GADA, and ODBASE*, volume 4275 of *Lecture Notes in Computer Science*. Springer Berlin Heidelberg, Berlin, Heidelberg, 2006.

[35] Nick Hawes. Building for the Future: Architectures for the Next Generation of Intelligent Robots. *Proceedings of a Symposium held in Honour of Aaron Sloman*, 2011.

[36] Nick Hawes, Aaron Sloman, and Jeremy Wyatt. Towards an integrated robot with multiple cognitive functions. *Proceedings of the Twenty-Second AAAI Conference on Artificial Intelligence (AAAI 2008), AAAI Press*, pages 1548–1553, 2008.

[37] Nick Hawes and Jeremy Wyatt. Engineering intelligent information-processing systems with CAST. *Advanced Engineering Informatics*, 24(1):27–39, 2010.

[38] Fredrik Heintz. *DyKnow: A Stream-Based Knowledge Processing Middleware Framework*. PhD thesis, Linköping Studies in Science and Technology. Dissertations #1240. Linköping University Electronic Press. 258 Pages., 2009.

[39] Fredrik Heintz. Semantically grounded stream reasoning integrated with ROS. *Intelligent Robots and Systems (IROS), 2013 IEEE/RSJ International Conference on*, pages 5935–5942, 2013.

[40] Fredrik Heintz, J Kvarnström, and Patrick Doherty. Stream-Based Reasoning Support for Autonomous Systems. *European Conference on Artificial Intelligence (ECAI)*, 2010.

[41] Fredrik Heintz, Jonas Kvarnstrom, and Patrick Doherty. A stream-based hierarchical anchoring framework. In *2009 IEEE/RSJ International Conference on Intelligent Robots and Systems*, pages 5254–5260. IEEE, October 2009.

[42] Fredrik Heintz, Jonas Kvarnström, and Patrick Doherty. Bridging the sense-reasoning gap: DyKnow - Stream-based middleware for knowledge processing. *Advanced Engineering Informatics*, 24(1):14–26, January 2010.

[43] Fredrik Heintz, Jonas Kvarnström, and Patrick Doherty. Stream-Based Hierarchical Anchoring. *KI - Künstliche Intelligenz*, 27(2):119–128, March 2013.

[44] Fredrik Heintz and D De Leng. Semantic information integration with transformations

for stream reasoning. *International Conference on Information Fusion (FUSION 2013)*, 2013.

[45] Ian Horrocks, Peter F. Patel-Schneider, Harold Boley, Said Tabet, Benjamin Grosof, and Mike Dean. SWRL: A Semantic Web Rule Language Combining OWL and RuleML. Technical report, W3C, 2004.

[46] Dominik Jain, Lorenz Mosenlechner, and Michael Beetz. Equipping robot control programs with first-order probabilistic reasoning capabilities. In *2009 IEEE International Conference on Robotics and Automation*, pages 3626–3631. IEEE, May 2009.

[47] Robert Kowalski and Marek Sergot. A logic-based calculus of events. In *Foundations of knowledge base management*, pages 23–55. Springer, 1989.

[48] Pat Langley, John E. Laird, and Seth Rogers. Cognitive architectures: Research issues and challenges. *Cognitive Systems Research*, 10(2):141–160, June 2009.

[49] Séverin Lemaignan. Grounding the Interaction: Knowledge Management for Interactive Robots. *PhD Thesis, Laboratoire d'Analyse et d'Architecture des Systèmes (CNRS) - Technische Universität München*, 2012.

[50] Séverin Lemaignan, Raquel Ros, E. Akin Sisbot, Rachid Alami, and Michael Beetz. Grounding the Interaction: Anchoring Situated Discourse in Everyday Human-Robot Interaction. *International Journal of Social Robotics*, 4(2):181–199, November 2011.

[51] Hector Levesque, Fiora Pirri, and Ray Reiter. Foundations for the situation calculus. *Linköping Electronic Articles in Computer and Information Science*, 3(18), 1998.

[52] Gi Hyun Lim, Il Hong Suh, and Hyowon Suh. Ontology-Based Unified Robot Knowledge for Service Robots in Indoor Environments. *IEEE Transactions on Systems, Man, and Cybernetics - Part A: Systems and Humans*, 41(3):492–509, May 2011.

[53] J. W. Lloyd. *Foundations of logic programming.* Springer-Verlag New York, Inc. New York, NY, USA, November 1984.

[54] I Lütkebohle. Facilitating re-use by design: A filtering, transformation, and selection architecture for robotic software systems. *ICRA'09 Workshop on Software Engineering for Robotics IV*, (section III), 2009.

[55] I Lütkebohle, R Philippsen, V Pradeep, E Marder-Eppstein, and S Wachsmuth. Generic middleware support for coordinating robot software components: The Task-State-Pattern. *Journal of Software Engineering for Robotics (JOSER)*, 2(1):20–39.

[56] Nikolaos Mavridis and Deb Roy. Grounded Situation Models for Robots: Where words and percepts meet. In *2006 IEEE/RSJ International Conference on Intelligent Robots and Systems*, pages 4690–4697. IEEE, October 2006.

[57] Sean Bechhofer Frank van Harmelen James Hendler Ian Horrocks Deborah L. McGuinness Peter F. Patel-Schneider Mike Dean, Guus Schreiber and Lynn Andrea Stein. OWL Web Ontology Language Reference. Technical report, W3C, 2004.

[58] Federico Pecora, Marcello Cirillo, Francesca Dell Osa, Jonas Ullberg, and Alessandro Saffiotti. A constraint-based approach for proactive, context-aware human support. *Journal of Ambient Intelligence and Smart Environments*, 4:347–367, 2012.

[59] Christian Peters, Thomas Hermann, and Sven Wachsmuth. User Behavior Recognition

For An Automatic Prompting System - A Structured Approach based on Task Analysis. *Proceedings of the 1st Int. Conf. on Pattern Recognition Applications and Methods (ICPRAM)*, 2:171, 2012.

[60] Axel Polleres, David Pearce, Stijn Heymans, and Edna Ruckhaus, editors. *Proceedings of the ICLP'07 Workshop on Applications of Logic Programming to the Web, Semantic Web and Semantic Web Services, ALPSWS 2007, Porto, Portugal, September 13th, 2007*, volume 287 of *CEUR Workshop Proceedings*. CEUR-WS.org, 2007.

[61] Morgan Quigley, Ken Conley, Brian P. Gerkey, Josh Faust, Tully Foote, Jeremy Leibs, Rob Wheeler, and Andrew Y. Ng. ROS: an open-source robot operating system. *Open Source Software Workshop of IEEE International Conference on Robotics and Automation (ICRA), 2009*, 2009.

[62] Surangika Ranathunga, Stephen Cranefield, and Martin Purvis. Identifying Events Taking Place in Second Life Virtual Environments. *Applied Artificial Intelligence*, 26(1-2):137–181, January 2012.

[63] Anand S Rao and Michael P Georgeff. Modeling Rational Agents within a BDI-Architecture. In James Allen, Richard Fikes, and Erik Sandewall, editors, *Proceedings of the 2nd International Conference on Principles of Knowledge Representation and Reasoning (KR'91)*, pages 473–484. Morgan Kaufmann publishers Inc.: San Mateo, CA, USA, 1991.

[64] Anand S. Rao and Michael P. Georgeff. BDI agents: From theory to practice. In *Proceedings of the first international conference on multi-agent systems (ICMAS-95)*, pages 312–319, 1995.

[65] C Bauckhage S. Wrede, M. Hanheide, Sagerer, and G. An active memory as a model for information fusion. *International Conference on Information Fusion, Stockholm, Sweden*, 1:198–205, 2004.

[66] L. Sabri, A. Chibani, Y. Amirat, and G. P. Zarri. Narrative reasoning for cognitive ubiquitous robots. In *IEEE/RSJ International Conference on Intelligent Robots and Systems (IROS 2011)*, 2011.

[67] Murray Shanahan. The event calculus explained. In *Artificial intelligence today*, pages 409–430. Springer, 1999.

[68] Evren Sirin, Bijan Parsia, Bernardo Cuenca Grau, Aditya Kalyanpur, and Yarden Katz. Pellet: A practical OWL-DL reasoner. *J. Web Sem.*, 5(2):51–53, 2007.

[69] Yale Song, David Demirdjian, and Randall Davis. Continuous body and hand gesture recognition for natural human-computer interaction. *ACM Transactions on Interactive Intelligent Systems*, 2(1):1–28, March 2012.

[70] Dennis Stachowicz and Geert-Jan M Kruijff. Episodic-Like Memory for Cognitive Robots. *IEEE Transactions on Autonomous Mental Development*, 4(1):1–16, March 2012.

[71] Mori Tenorth and Michael Beetz. KNOWROB - knowledge processing for autonomous personal robots. In *2009 IEEE/RSJ International Conference on Intelligent Robots and Systems*, pages 4261–4266. IEEE, October 2009.

[72] Moritz Tenorth and Michael Beetz. Knowledge Processing for Autonomous Robot Control. *Proceedings of the AAAI Spring Symposium on Designing Intelligent Robots: Reintegrating AI. Stanford, CA: AAAI Press, 2012*, 2012.

[73] Moritz Tenorth, Alexander Clifford Perzylo, Reinhard Lafrenz, and Michael Beetz. The RoboEarth language: Representing and exchanging knowledge about actions, objects, and environments. *2012 IEEE International Conference on Robotics and Automation*, (3):1284–1289, May 2012.

[74] André Ückermann, Robert Haschke, and Helge Ritter. Real-Time 3D Segmentation of Cluttered Scenes for Robot Grasping. *IEEE-RAS International Conference on Humanoid Robots (Humanoids 2012), Osaka, Japan*, 2012.

[75] Vangelis Vassiliadis, Jan Wielemaker, and Chris Mungall. Processing OWL2 ontologies using Thea: An application of logic programming. In *OWLED*, volume 529, 2009.

[76] V Verma and A Jónsson. Universal executive and PLEXIL: Engine and language for robust spacecraft control and operations. *American Institute of Aeronautics and Astronautics Space Conference*, pages 1–19, 2006.

[77] David E. Watson. Book review: Blackboard Architectures and Applications Edited by V. Jagannathan, Rajendra Dodhiawala, and Lawrence S. Baum (Academic Press). *ACM SIGART Bulletin*, 1(3):19–20, October 1990.

[78] Jan Wielemaker, Tom Schrijvers, Markus Triska, and Torbjörn Lager. SWI-Prolog. *Theory and Practice of Logic Programming*, 12(1-2):67–96, 2012.

[79] R. Wood, P. Baxter, and T. Belpaeme. A review of long-term memory in natural and synthetic systems. *Adaptive Behavior*, 20(2):81–103, December 2011.

[80] S Wrede. An information-driven architecture for cognitive systems research. *Ph.D. dissertation, Faculty of Technology - Bielefeld University*, 2009.

[81] Pouyan Ziafati, Mehdi Dastani, John-Jules Meyer, and Leendert van der Torre. Agent Programming Languages Requirements for Programming Autonomous Robots. *ProMAS 2012, Springer, Heidelberg*, LNAI 7837:35–53, 2013.

[82] Pouyan Ziafati, Mehdi Dastani, John-Jules Meyer, and Leendert van der Torre. Event-Processing in Autonomous Robot Programming. *Proceedings of the 12th International Conference on Autonomous Agents and Multiagent Systems*, pages 95–102, 2013.

[83] Pouyan Ziafati, Yehia Elrakaiby, Mehdi Dastani, Leendert van der Torre, Marc van Zee, John-Jules Meyer, and Holger Voos. Reasoning on Robot Knowledge from Discrete and Asynchronous Observations. *AAAI Spring Symposium on Knowledge Representation and Reasoning in Robotics, Stanford, 2014*, 2014.

[84] Pouyan Ziafati, Fulvio Mastrogiovanni, and Antonio Sgorbissa. Fast Prototyping and Deployment of Context-Aware Smart Outdoor Environments. *2011 Seventh International Conference on Intelligent Environments*, pages 206–213, July 2011.

Received June 2015

Going Forth and Drawing Back: An Intensional Approach in Nonmonotonic Inference

Yi Mao
atsec Information Security Corporation
9130 Jollyville Road, Suite 260, Austin, TX78759, USA

Beihai Zhou
Department of Philosophy, Peking University Beijing, 100871, P.R. China

Beishui Liao*
Center for the Study of Language and Cognition/Department of Philosophy, Zhejiang University, Hangzhou, 310028, P.R. China
University of Luxembourg, L-1359, Luxembourg
baiseliao@zju.edu.cn

Abstract

We decompose a nonmonotonic inference $\Gamma \mid\!\sim \alpha$ into two stages: going forth to deduce all default conclusions in a logic system named DC (short for Default Conclusions), and drawing back less preferable conclusions in the face of conflicting default conclusions, based on a binary "more preferable" relation defined on those subformulas of Γ that are also its deductive consequences. Under the possible-world semantics framework, we construct a set selection function and build up a theory of semantics, as a variant of traditional selection function semantics. We prove that the underlying logic system DC is sound and complete with respect to set selection function semantics. Using this two-layer mechanism, we account for benchmark examples including the Nixon Diamond and the Penguin Principle.

Keywords: nonmonotonic inference, defaults, possible world semantics,

The research reported in this article was financially supported by the National Social Science Foundation Major Project of China under grant No.11&ZD088 and No.12&ZD119, and the National Natural Science Foundation of China under grant No.61175058.

*Corresponding author.

1 Introduction

Our knowledge about the world in which we live consists mostly of statements like "Birds fly", "Potatoes contain vitamin C", etc. These are often called default statements or defaults. People observe regularities in the world and codify them in defaults to express their law-like nature. The defaults enable people to predict what the future is about to bring based on the observed regularities, and then people act on these predictions accordingly. For instance, suppose that Tweety is a bird. Since we know that birds fly, we predict that Tweety flies. In order to prevent Tweety from flying away, the cage to keep it should have a lid. The reasoning involved to infer "Tweety flies" from "Birds fly" and "Tweety is a bird" is usually called default reasoning or defeasible reasoning. The term "defeasible reasoning" indicates that any conclusions derived from defaults can be invalidated by providing new evidence. In the example, if we later learn that Tweety is a penguin, then we should withdraw the previous conclusion and claim instead that Tweety does not fly. Now, it is not necessary for Tweety's cage to have a lid. As the growth of the premises may cause the retraction of previously drawn conclusions, nonmonotonicity is intimately connected with defeasibility. This type of reasoning only warrants conclusions, but does not guarantee their truth. Researchers have produced intensive work in this area, although their approaches vary. We shall very briefly review some major existing theories of defeasible reasoning and position our theory in the coarse road map of this research area.

Regardless of the diversity of approaches that researchers take to characterize reasoning with defaults, they all aim to account for the following benchmark examples:

(1) If it rains, the ground gets wet. It rains. / The ground gets wet.

(2) Birds fly. Tweety is a bird. / Tweety flies.

(3) Birds fly. Tweety is a bird. Tweety does not fly. / Tweety does not fly.

(4) Whales are mammals. Marine creatures normally are not mammals. Willy is a whale. Willy is a marine creature. / Willy is a mammal.

(5) If it rains, the ground gets wet. It rains and the wind blows. / The ground gets wet.

(6) Quakers are pacifists. Republicans are not pacifists. Nixon is a Quaker. / Nixon is a pacifist.

(7) Quakers are pacifists. Republicans are not pacifists. Nixon is a Quaker. Nixon is a republican. / Nixon is a pacifist?? Nixon is not a pacifist??

(8) Birds fly. Penguins do not fly. Tweety is a bird. Tweety is a penguin. / Tweety flies?? Tweety does not fly??

(9) Penguins are birds. Birds fly. Penguins do not fly. Tweety is a bird. Tweety is a penguin. / Tweety does not fly.

(10) College students are adults. Adults can drive. John is a college student. / John can drive.

(11) College students are adults. Adults are employed. John is a college student. / John is employed ??

In the above examples, sentences appearing before the slash symbol "/" are premises, and those after the slash marks are conclusions. Conclusions marked by "??" are not commonly accepted. (1) and (2) show that people use default *Modus Ponens* to detach the consequents of defaults. Let us call the obtained conclusions *default conclusions.* (3) indicates that factual statements have higher priority to be conclusions than those obtained by applying default *Modus Ponens.* We call this the *fact-first* principle. (4) reflects the intuition that deductively-derived conclusions are more trustworthy than default conclusions. Default conclusions should always yield to deductive ones in case of conflict. The application of default *Modus Ponens* in (5) and (6) are not affected by some irrelevant fact(s) or default(s). The premises in (7) are not contradictory, but putting them together causes incompatible conclusions, none of which are acceptable for the entire set of premises. This inference pattern is known as the Nixon Diamond. (8) is in the same form as (7). One who has strong inclination to use an additional implicit premise "penguins are birds" could actually conduct the reasoning as in (9). The increase of premises from (6) to (7) has blocked a previously-drawn conclusion, while adding one more premise from (8) to (9) helps to lead to a conclusion that would otherwise not be possible. (9) is known as the Penguin Principle. It shows that each premise in a default reasoning does not weight equally. People tend to give higher priority to premises carrying more specific information and prefer arguments based on the most specific defaults. This tendency is called *specificity.* In artificial intelligence, the specificity principle is regarded by many as a very important general principle of commonsense reasoning. (10) and (11) suggest that pointwise transitivity should not be completely abandoned, nor should it be accepted without any restriction.

Based upon the examples analyzed, we summarize five distinct but closely related basic features regarding default reasoning.

First, it is not required to derive true conclusions from true premises. Instead, it is only expected to derive acceptable conclusions from true premises. Unlike conclusions drawn from the classical logic, conclusions from default reasoning can be revised or retracted in the face of new information.

Second, being nonmonotonic to the increase of premises, a conclusion deduced from a part of premises may no longer be a conclusion of all premises. All premises must be taken into account when a conclusion of a default reasoning is under examination. It is very important to respect this feature when we deal with nonmonotonic reasoning. Hidden premises should not be allowed, as the conclusions may differ with or without them. Those who tend to conclude that Tweety does not fly from example (8) are actually conducting an inference based on (9). If they use pattern (8) to express their premises, they have hidden the premise "Penguins are birds", which is the trigger to invoke the specificity. Without this information, the inference falls into the pattern of the Nixon Diamond, from which nothing can be concluded.

Third, it could be the case that all of the premises are compatible, but different parts of the premises may imply contradictory conclusions. As contradiction is not acceptable, contradictory conclusions from some parts of the premises (let us call them *local conclusions*) cannot sustain to be conclusions for the entire premise set (let us call them *global conclusions*).

Fourth, certain tendencies such as fact-first principle and specificity hidden in background knowledge tip the balance between contradictory local conclusions and determine which local conclusions can be global conclusions for a given premise set, and which ones must be given up.

Finally, corresponding to two types of conclusions (i.e., local vs. global), there are two kinds of inferences. One infers a local conclusion from a certain part of the premises, while the other infers a global conclusion from the entire premise set. To differentiate them, we call the first kind of inference *local inference*, and the second one *global inference*. When the inference in question is monotonic, these two notions coincide.

Among various theories, there are basically two approaches to formulate defaults. One approach is like Reiter's default logic ([28]), in which defaults are expressed as domain-specific inference rules. These rules have the form $\frac{A:B}{C}$, where A, B, and C are ordinary first-order formulas. "Bird fly" is informally read as "If it is provable that x is a bird, and it is not provable that x does not fly, then we may infer that x flies". We can apply the default rule to Tweety and derive that Tweety flies. The defeasibility stems from the consistency check for applying the default rule. With more premises, that Tweety does not fly may become provable and the default rule fails to apply. It is worth noting that in this approach, what is provable is determined by what is not provable. Provability is defined in terms of a fixed-point, to avoid circularity.

The other approach is to formalize defaults as statements, rather than rules. This again can be classified into two subcategories. Theories like preferential entailment (e.g., [29, 16]) and circumscription (e.g., [21, 17]) deploy material implications to express defaults. To reflect that the consequent of a default can only be defeasibly derived from its antecedent, an extra "normality condition" is added into the antecedent of a default as in $\forall x(bird(x) \wedge \neg ab(x) \rightarrow fly(x))$. It is assumed that things are as normal as possible, unless otherwise stated. The inference is based on first-order logic and *Modus Ponens* in particular. The defeasibility of this kind of inference can be shown by adding a later discovery that something in question is not as normal as assumed, e.g., "Tweety is abnormal", to the premises.

The intensional approach (e.g., [8, 9, 3, 23, 2]) uses conditionals in the form $\alpha > \beta$ as the logical representation of defaults. The defeasibility of defaults is intrinsic to the modal operator $>$, which is interpreted in a possible-world semantics via a selection function that selects normal worlds for truth evaluation of defaults. Roughly, $\alpha > \beta$ is true if β is true in all normal worlds where α is true. The *Modus Ponens* for $>$ (i.e., $\alpha \wedge (\alpha > \beta) \rightarrow \beta$) is not valid for such conditionals, because the actual world may not always be a normal world. Deducing β from $\alpha \wedge (\alpha > \beta)$ depends on the assumption that the actual world is as normal as possible.

These approaches demand a consistency check before applying a default rule or *Modus Ponens* to *infer* a certain conclusion, to ensure that the opposite conclusion cannot be *inferred*. The first occurrence of "infer" is a *global inference* that infers a conclusion from the entire premise set, while the second one is a *local inference* that infers the opposite conclusion from some premises. To block potential opposite conclusions that might otherwise be locally inferred, all premises must be taken into consideration against the consistency check. In nonmonotonic reasoning, some local conclusions are admitted to be global conclusions only if they pass the consistency check. Though conducting consistency checks and taking all premises into account during every step of inference are theoretically possible, to carry out this task is computationally expensive.

We intend to propose a solution based on the idea of conducting nonmonotonic inference in two layers. Rather than handling inconsitency at each time to apply a certain inference rule (domain-dependent default rule or classical *modus ponens*), we my delay the inconsistency handling to the very end when partially inferable conclusions are all derived. The defeasible inference is broken down into two phases. The first phase allows to get all local conclusions. This phase of inference is supported by a logic which is similar to the classical logic in spirit. Default conclusions could well be contradictory. For instance, we get "Tweety flies" from "Tweety is a bird" and "Birds fly". Also, we may get "Tweety does not fly" from "Tweety is a penguin" and "Penguins do not fly". The local inference captured by this logic is actually monotonic. The second phase is to deal with inconsistent local conclusions. A local conclusion may not be a global conclusion of the entire premise set. Some standards like the specificity rule that people tend to follow in commonsense reasoning could tip the balance between conflicting local conclusions in favour of, say, the conclusion that Tweety does not fly as a penguin. The overall inference will be nonmonotonic. The proposed solution defers the management of inconsistency to the very end, when local conclusions are all deduced, rather than handling inconsistency at each application of a certain inference rule (domain-dependent default rule or classical *Modus Ponens*).

For the purpose of obtaining all local conclusions during the first phase, we develop a formal logic system DC (short for Default Conclusions), which particularly facilitates the deduction of default conclusions. We formalize the defaults as statements. Furthermore, default statements are not represented by material implications but by default implications to reflect their intensionality. We made these two decisions based on the reasons argued by Pelletier and Asher ([26]).

The major difference between our system DC and other intensional approach systems (e.g., [8, 9, 3, 23, 2]) is that we make a weak version of *Modus Ponens* $(\alpha \wedge (\alpha > \beta)) > \beta$ to be an axiom. We name it *Default Modus Ponens.* It is the core of the inference engine that detaches the consequents of defaults and "produces" default conclusions. In the classical logic, this follows so naturally that any system validates *Modus Ponens.* It is the backbone on which a monotonic inference relies. Unfortunately, the Default *Modus Ponens* is not similarly widely accepted and has not yet drawn enough attention. The Default *Modus Ponens* is not provable in Delgrande's systems, nor is it validated by the semantics of Asher and his coauthors.

Without Default *Modus Ponens*, the consequents of defaults are indirectly obtainable under classical *Modus Ponens* via various forms of transformation ([2]) or extension ([9]) or normalization ([3]). The idea behind these sophisticated procedures is rather simple — treating the default implication $\alpha > \beta$ as if the material implication $\alpha \to \beta$ whenever possible. However, checking the satisfaction of the condition "whenever possible" amounts to a consistency check, which is not so simple. Their formal logic systems are mostly used as background logics against which consistency checking can be done. With Default *Modus Ponens*, default conclusions can now be directly inferred. The complexity of managing the assumption of normality or assumption of relevance as in [9] is avoided.

Although it is tempting to accept Default *Modus Ponens* and let it lead to default conclusions in nonmonotonic inference, there are technical difficulties that must be addressed. As Default *Modus Ponens* has the embedded $>$ operator, its characterizing frame condition looks cumbersome in the widely-accepted modal conditional semantics (e.g., [24, 25]) and can hardly make any use of it. We not only argue that the Default *Modus Ponens* should play an equally crucial role in nonmonotonic reasoning as *Modus Ponens* in monotonic reasoning, but also improve the semantics by lifting the selection function to a higher level so that we overcome the characterization difficulty. The details of the intuition towards the acceptance of Default *Modus Ponens* and its semantic characterization will be discussed in later sections.

Our work presented in this paper is twofold: First, we provide an underlying logic in which all default conclusions of given premises can be derived. Second, we provide a method to handle inconsistent default conclusions by comparing the strength of the premise sets that support them. Our goal is to show that the two-phase intensional approach not only complements existing theories by providing a new perspective to tackle the problem, it also offers advantages.

2 The Logic DC of Default Conclusions

2.1 Logical Preliminaries

The language $\mathfrak{L}_>$ contains a denumerable set of propositional variables $P = \{p_0, p_1, \ldots\}$, truth-functional connectives $\neg$ and $\rightarrow$, and a binary operator $>$. Formulas are defined as usual. p, q, r, etc. will be used for atomic sentences, α, β, γ etc. for formulas, Fla for the set of all formulas, Γ, Δ, Θ, etc. for sets of formulas. $\vee$, $\wedge$, $\leftrightarrow$ and $\not>$ are introduced as abbreviations. $\alpha \not> \beta$ abbreviates $\neg(\alpha > \beta)$.

Precedence order is assumed from the strongest to the weakest among connectives like this: $\neg$, $\wedge$, $\vee$, $>$, $\rightarrow$, $\leftrightarrow$. For instance, $\alpha > \beta \wedge \gamma \rightarrow \alpha > \beta$ should be read as $(\alpha > (\beta \wedge \gamma)) \rightarrow (\alpha > \beta)$. If the context is not ambiguous, some parentheses may be omitted. Note that $>$ having a higher precedence than $\rightarrow$ should not be understood as the $>$ operator is stronger than the material implication $\rightarrow$ in the sense that $(\alpha > \beta) \rightarrow (\alpha \rightarrow \beta)$. In the system DC, neither $(\alpha > \beta) \rightarrow (\alpha \rightarrow \beta)$ nor $(\alpha \rightarrow \beta) \rightarrow (\alpha > \beta)$ is valid.

2.2 Key Axioms of System DC

Before we formally lay out the system DC, we give some explanations on the selection of its key axioms.

2.2.1 *Default Modus Ponens*

In the underlying logic, we expect to get all default conclusions. Ideally, a formula ϕ is a default conclusion of a finite premise set Γ iff $\bigwedge\Gamma > \phi$ is provable in the system. Since at this level, we do not plan to handle conflicting default conclusions, "being a default conclusion" is actually a monotonic relation in the sense that if ϕ is a default conclusion of Γ then ϕ is a default conclusion of all supersets of Γ. The core of inferring default conclusions from premises containing defaults is how to detach the consequents of defaults. In the classical logic, formula $(\alpha \wedge (\alpha \rightarrow \beta)) \rightarrow \beta$ describes the *Modus Ponens* for the material implication $\rightarrow$. This does not apply to defaults, as we formulate them as conditionals in the form of $\alpha > \beta$ rather than $\alpha \rightarrow \beta$. A simple-minded transplantation of *Modus Ponens* to defaults like $(\alpha \wedge (\alpha > \beta)) \rightarrow \beta$ does not work, because this formula contradicts the exception tolerance feature of the default $\alpha > \beta$. For example, it is compatible to claim that Tweety is a bird and birds fly, but that Tweety does not fly. This is to say that $\{\alpha > \beta, \alpha, \neg\beta\}$ is consistent. But $(\alpha \wedge (\alpha > \beta)) \rightarrow \beta$ leaves no room for this kind of exceptional case.

Considering that $\alpha > \beta$ is read as "if α then normally β", the operator $>$ is to capture the "normally follow" relationship. Suppose that we accept the default if α then normally β, and further suppose that we are in a situation where α is true. Then, normally, β should be true under this situation. To formulate what has just been said, we have $(\alpha \wedge (\alpha > \beta)) > \beta$, which delineates a similar *Modus Ponens* for the $>$ operator. As we mentioned earlier, we label this formula as Default *Modus Ponens*. Boutilier ([7]) proves it as a theorem of the system C4TO, but we will take it as a very basic axiom of the system DC that is designed to obtain default conclusions. The Default *Modus Ponens* plays an important role to detach the consequent of a default when its antecedent is present. It makes the defaults participate in the inference of obtaining default conclusions in a rather direct way, but it does not exaggerate the certainty of these conclusions as $(\alpha \wedge (\alpha > \beta)) \rightarrow \beta$ would have done. Using the $>$ operator as the main connective in the Default *Modus Ponens* indicates that they are default conclusions. This differentiates them from other conclusions that can be classically implied from the antecedent, say, $(\alpha \wedge (\alpha > \beta)) \rightarrow \alpha$.

2.2.2 Restricted Pointwise Transitivity

Let us look at two examples:

(1a)	If the ground gets wet, then roads are slippery.
(1b)	If it rains, then the ground gets wet.
(1c)	It is raining.
(1d)	Roads are slippery.

(2a)	If it rains, then crop harvest is expected.
(2b)	If hurricane comes, then it rains.
(2c)	Hurricane comes.
(2d)	Crop harvest is expected.

Both examples illustrate the use of pointwise transitivity:

PTRAN $(\alpha \wedge (\alpha > \beta) \wedge (\beta > \gamma)) > \gamma$

However, there is a difference between the examples. We can accept conclusion (1d) inferred from (1a)-(1c), but we tend to object obtaining (2d) from (2a)-(2c). Before we hear (2b), we may accept (2a) without any reservation. After we hear (2b), we are inclined to rescind our approval to (2a). The conditions we use to evaluate (2a) before we hear (2b) seem not to be the same as those we use to evaluate (2a) after we hear (2b). Once we hear (2b), we begin to think in terms of rain caused by hurricane, and we see that the argument explicitly requires abnormal rain cases that disjoint the normal rain cases considered in (2a). This explains our change of mind concerning the acceptance of (2a). Instead of (2a), what is actually used in the argument is (2a'):

(2a') If hurricane comes and it rains, then crop harvest is expected.

(2a') is a false statement, and thus the argument fails. The success of the argument expressed in example 1 is due to the truth of (1a'):

(1a') If it rains and the ground gets wet, then roads are slippery.

The formalization of unrestricted pointwise transitivity PTRAN does not reflect implicitly-used premises (1a') and (2a'), and will mistakenly admit the argument for (2d). Thus, we reject unrestricted PTRAN in favor of restricted pointwise transitivity RPT:

RPT $((\alpha \wedge \beta) > \gamma) \rightarrow (\alpha \wedge (\alpha > \beta) \wedge (\beta > \gamma)) > \gamma$

RPT precisely differentiates example 1 from example 2. While example 1 is a valid argument justified by RPT, example 2 is blocked. We want our system to support RPT, and take it as an axiom.

Transitivity is also often expressed in this version:

TRAN $((\alpha > \beta) \wedge (\beta > \gamma)) \rightarrow (\alpha > \gamma)$

The following invalid argument shows that TRAN fails, since (3c) is false.

(3a)	College students are adults.
(3b)	Adults are employed.
(3c)	College students are employed.

However, we do quite often deploy TRAN in default reasoning:

(4a)	College students are adults.
(4b)	Adults know how to drive.
(4c)	College students know how to drive.

In the first example, normal conditions we employ in evaluating (3a) are not the same as conditions we employ in evaluating (3b). Although college students are adults, they are exceptional adults with regard to consideration of their status of employment. It seems that we evaluated the wrong conditional when we evaluate (3b). What really should be evaluated is the new premise:

(3b') Adult college students are employed.

However, this new premise is false. Thus, the conclusion (3c) is not acceptable.

In the second example, normal conditions where college students are adults are also normal conditions where adults know how to drive. Regarding driving capability, college students are as normal as other classes of adults. The new premise (4b') that is really evaluated is true in this case.

(4b') Adult college students know how to drive.

This might explain why we are inclined to approve the transitivity in example 4, but rescind such an approval in example 3. It appears that a certain degree of transitivity in default reasoning should be granted, but something like example 3 should not be allowed. For this purpose, Nute proposed restricted transitivity:

RT $((\alpha \wedge \beta) > \gamma) \rightarrow ((\alpha > \beta) \rightarrow (\alpha > \gamma))$

It seems that our system should support RT as well. If not, it should at least support a weaker version RT$'$:

RT$'$ $((\alpha \wedge \beta) > \gamma) \rightarrow ((\alpha > \beta) \wedge (\beta > \gamma) \rightarrow (\alpha > \gamma))$

Nevertheless, our system supports neither RT nor RT$'$. Taking RT or RT$'$ as an axiom would imply PTRAN in our system, which we would like to avoid. We decide not to include RT or RT$'$. The strong side of our system is that it gives a good capture of Default *Modus Ponens*. RPT can be viewed as transitive *Modus Ponens*. It shows how Default *Modus Ponens* interacts with transitivity. We take it as an axiom to complete our characterization of Default *Modus Ponens*. The weak side of our system is that it leaves out the consideration of restricted transitivity.

2.3 The System DC

The system DC is designed to catch all default conclusions that "normally follow" from the antecedents of defaults. It is the smallest logic that contains the propositional calculus (PC) and is closed under the following axiom schemata and rules of inference.

Axiom schemata:

CK	$(\alpha > (\beta \to \gamma)) \to ((\alpha > \beta) \to (\alpha > \gamma))$
DMP	$(\alpha \wedge (\alpha > \beta)) > \beta$
ID	$\alpha > \alpha$
RPT	$((\alpha \wedge \beta) > \gamma) \to ((\alpha \wedge (\alpha > \beta) \wedge (\beta > \gamma)) > \gamma)$

Axiom CK in system DC is in parallel to axiom K in modal logic. As we take the conditional modal logic approach to deal with defaults, we would like to have it in our system. The name "CK" indicates that it is a variant of axiom K in conditional logic setting. $\alpha > \alpha$ asserts that every premise is also a default conclusion of itself. It is widely accepted by many authors (e.g., Delgrande, Asher etc.) as an axiom known as *identity* axiom. We adopt this axiom and use a short name "ID" to refer to it.

Rules of inference:

MP	From α and $\alpha \to \beta$, infer β
REQ	From $\beta \leftrightarrow \gamma$ and α, infer $\alpha[\gamma/\beta]$, where β is a subformula of α
RN	From β, infer $\alpha > \beta$
RM	From $\alpha > \beta$, infer $(\alpha \wedge \gamma) > \beta$

All these rules of inferences except RM sound familiar to a modal logic system. RM is designated to deal with augmented additional information towards defaults that have been established as theorems of DC. We know that the "normally follow" relationship $>$ is not monotonic to its antecedent. That is, MON below should not be a valid form:

MON $\quad (\alpha > \beta) \to ((\alpha \wedge \gamma) > \beta)$

RM is much weaker than MON. Only for those $(\alpha > \beta)$ that are provable in DC, adding more information γ will have no impact on the derivation of default conclusion β. From an example, as $(\alpha \wedge (\alpha > \beta)) > \beta$ is an axiom of DC, it seems reasonable to accept $(\gamma \wedge \alpha \wedge (\alpha > \beta)) > \beta$ in DC to show that the default conclusions are monotonic to additional information. In case where γ is actually $\neg\beta$, which causes inconsistency, the conflicting default conclusions will be resolved in the next stage. β will be blocked, and $\neg\beta$ will be the global conclusion instead. Because system DC is intended to provide us all possible default conclusions, we take RM as an inference rule of DC. The consideration of consistency check-up among candidate default conclusions is left to the second stage in our two-phase process.

With such a specified system DC, we can get the following derived rules:

RIC	From $\alpha \to (\beta > \gamma)$, infer $(\alpha \wedge \beta) > \gamma$
RCK	From $\beta_1 \wedge ... \wedge \beta_n \to \beta$, infer $(\alpha > \beta_1) \wedge ... \wedge (\alpha > \beta_n) \to (\alpha > \beta)$, for any $n \geqslant 1$
RAM	From $\alpha \to \beta$ and $\beta > \gamma$, infer $\alpha > \gamma$
RIN	From $\alpha \to \beta$ infer $\alpha > \beta$

Proof. For RIC, suppose that $\vdash \alpha \to (\beta > \gamma)$. Then $\vdash \alpha \leftrightarrow (\alpha \wedge (\beta > \gamma))$, and also $\vdash (\alpha \wedge \beta) \leftrightarrow (\alpha \wedge \beta \wedge (\beta > \gamma))$, by PC. By axiom DMP, we have $\vdash (\beta \wedge (\beta > \gamma)) > \gamma$. By RM, $\vdash (\alpha \wedge (\beta \wedge (\beta > \gamma))) > \gamma$. Then by REQ, $\vdash (\alpha \wedge \beta) > \gamma$.

For RCK, suppose that $\vdash (\beta_1 \wedge ... \wedge \beta_n) \to \beta$. By RN, we have $\vdash \alpha > (\beta_1 \wedge ... \wedge \beta_n \to \beta)$. By CK and MP, we have $\vdash (\alpha > (\beta_1 \wedge ... \wedge \beta_n)) \to (\alpha > \beta)$. By CC (proved in a theorem below), we have $\vdash (\alpha > \beta_1) \wedge ... \wedge (\alpha > \beta_n) \to (\alpha > (\beta_1 \wedge ... \wedge \beta_n))$. Hence, by PC, we have $\vdash (\alpha > \beta_1) \wedge ... \wedge (\alpha > \beta_n) \to (\alpha > \beta)$.

For RAM, suppose that $\vdash \alpha \to \beta$ and $\vdash \beta > \gamma$. By RM, $\vdash (\alpha \wedge \beta) > \gamma$. By REQ, $\vdash (\alpha \wedge (\alpha \to \beta)) > \gamma$. Since $\vdash \alpha \to \beta$, $\vdash (\alpha \wedge (\alpha \to \beta)) \leftrightarrow \alpha$. By REQ, $\vdash \alpha > \gamma$.

For RIN, by rule RCK, we can, from $\alpha \to \beta$, infer $(\alpha > \alpha) \to (\alpha > \beta)$. Since $\alpha > \alpha$ is an axiom, we have $\alpha > \beta$. □

Theorem 1. *The following formulas are theorems of the system DC.*

Th$_{DC}$1 (CR) $(\alpha > (\beta \wedge \gamma)) \to ((\alpha > \beta) \wedge (\alpha > \gamma))$

Th$_{DC}$2 (CC) $((\alpha > \beta) \wedge (\alpha > \gamma)) \to (\alpha > (\beta \wedge \gamma))$

Th$_{DC}$3 $(\gamma \wedge \alpha \wedge (\alpha > \beta)) > \beta$

Th$_{DC}$4 $(((\alpha \wedge \beta) > \gamma) \wedge \alpha \wedge (\alpha > \beta) \wedge (\beta > \gamma)) > \gamma$

Proof. Th$_{DC}$1:

(1)	$(\beta \wedge \gamma) \to \beta$	PC
(2)	$\alpha > ((\beta \wedge \gamma) \to \beta)$	(1),RN
(3)	$(\alpha > (\beta \wedge \gamma)) \to (\alpha > \beta)$	(2),CK,MP
(4)	$(\beta \wedge \gamma) \to \gamma$	PC
(5)	$\alpha > (\beta \wedge \gamma \to \gamma)$	(4),RN
(6)	$(\alpha > (\beta \wedge \gamma)) \to (\alpha > \gamma)$	(5),CK,MP
(7)	$(\alpha > (\beta \wedge \gamma)) \to ((\alpha > \beta) \wedge (\alpha > \gamma))$	(3),(6),PC

Proof. $\mathsf{Th}_{DC}2$:

(1)	$\beta \rightarrow (\gamma \rightarrow (\beta \wedge \gamma))$	PC
(2)	$\alpha > (\beta \rightarrow (\gamma \rightarrow (\beta \wedge \gamma)))$	(1),RN
(3)	$(\alpha > \beta) \rightarrow \alpha > (\gamma \rightarrow (\beta \wedge \gamma))$	(2),CK,MP
(4)	$(\alpha > (\gamma \rightarrow (\beta \wedge \gamma))) \rightarrow ((\alpha > \gamma) \rightarrow \alpha > (\beta \wedge \gamma)))$	CK
(5)	$(\alpha > \beta) \rightarrow ((\alpha > \gamma) \rightarrow \alpha > (\beta \wedge \gamma))$	(3),(4),PC
(6)	$((\alpha > \beta) \wedge (\alpha > \gamma)) \rightarrow (\alpha > (\beta \wedge \gamma))$	(5),PC

$\mathsf{Th}_{DC}3$ follows from axiom DMP and rule RM. $\mathsf{Th}_{DC}4$ can be obtained from axiom RPT and rule RIC.

In system DC, we can prove PTRAN from RT or RT′, as we discussed in section 2.2.2. We repeat them below:

RT	$((\alpha \wedge \beta) > \gamma) \rightarrow ((\alpha > \beta) \rightarrow (\alpha > \gamma))$
RT′	$((\alpha \wedge \beta) > \gamma)) \rightarrow ((\alpha > \beta) \wedge (\beta > \gamma) \rightarrow (\alpha > \gamma))$
PTRAN	$(\alpha \wedge (\alpha > \beta) \wedge (\beta > \gamma)) > \gamma$

From RT, we can get RTRAN by RM and MP:

RTRAN If $\vdash \beta > \gamma$, then $\vdash (\alpha > \beta) \rightarrow (\alpha > \gamma)$

From RTRAN, we can get PTRAN as shown below:

Proof.

(1)	$(\beta \wedge (\beta > \gamma)) > \gamma$	DMP
(2)	$((\alpha \wedge (\alpha > \beta) \wedge (\beta > \gamma)) > (\beta \wedge (\beta > \gamma))) \rightarrow$ $((\alpha \wedge (\alpha > \beta) \wedge (\beta > \gamma)) > \gamma)$	(1),RTRAN
(3)	$(\alpha \wedge (\alpha > \beta)) > \beta$	DMP
(4)	$(\alpha \wedge (\alpha > \beta) \wedge (\beta > \gamma)) > \beta$	(3),RM
(5)	$(\beta > \gamma) > (\beta > \gamma)$	ID
(6)	$(\alpha \wedge (\alpha > \beta) \wedge (\beta > \gamma)) > (\beta > \gamma)$	(5),RM
(7)	$(\alpha \wedge (\alpha > \beta) \wedge (\beta > \gamma)) > (\beta \wedge (\beta > \gamma))$	(5),(6),CC
(8)	$((\alpha \wedge (\alpha > \beta) \wedge (\beta > \gamma)) > \gamma)$	(2),(7),MP

FromRT′, we can get RTRAN′:

RTRAN′ If $\vdash \beta > \gamma$, then $\vdash (\alpha > \beta) \wedge (\beta > \gamma) \rightarrow (\alpha > \gamma)$

If $\vdash \beta > \gamma$, then $\vdash (\alpha > \beta) \leftrightarrow (\alpha > \beta) \wedge (\beta > \gamma)$. So, RTRAN′ is equivalent to RTRAN. Thus, PTRAN can also be derived from RT′ via RTRAN.

Based on the several formally-proved theorems of the system DC, we have explored a variety form of transitivity of the $>$ operator and their interaction with

DMP. As we design our system with a focus on the detachment of a default consequent from a default, we leave RT and RT′ out of our consideration in favor of RPT.

From the proofs that we have constructed, DMP is frequently used to show some properties of $>$. This is one advantage of introducing DMP as an axiom. We can now study the properties of $>$ as an operator that is independent of the $\rightarrow$ operator. For us, the $>$ operator is not only important to represent defaults, but also equally important to capture the "normally follow" behavior of these defaults appearing in inferences. The $\rightarrow$ operator in the classical logic is used to formalize the material implication "if...then" on the one hand, and also to mirror our monotonic inference on the other hand. These two things are bridged via the deduction theorem in the classical logic. On the contrary, the $>$ operator has long been used only for representing defaults in the intentional approach. As far as the inference is concerned, the formulas connected by the $>$ operator like $\alpha > \beta$ are transformed or converted to something more or less like $\alpha \rightarrow \beta$, in order to make the classical *Modus Ponens* applicable. At the end, the inferences with the $>$-formulas involved are reduced to applying *Modus Ponens*. The $>$ operator was rarely linked to representing features of nonmonotonicity that stem from "normally follow" in a direct manner. For instance, we may ask a question like this: will there be some logic system $S(L)$ about $>$ such that $\vdash_{S(L)} \alpha > \beta$ iff $\alpha \mathrel{|\!\sim}_{S(L)} \beta$. Although results in Mao ([19]) indicate that the link between the representation of defaults and inference features of $>$ is not as straightforward as it is for the material implication $\rightarrow$, DMP is the first step to making that link, and it frees inferences with defaults from being peripheral to classical ones.

3 Semantics

3.1 Set Selection Function

Since we take $(\alpha \wedge (\alpha > \beta)) > \beta$ as an axiom to grant the detachment of the consequents of $>$-formulas, the semantics to be developed should validate it. Let us start with the existing selection function semantics for conditional logics that is reused by Delgrande, Asher and Morreau for default statements. Think back to what the selection function provides: if $\alpha > \beta$ is true at w, then $*(w, \|\alpha\|^{\mathfrak{M}}) \subseteq \|\beta\|^{\mathfrak{M}}$[1], and vice versa. This applies to all possible worlds. Choosing another possible world w', again, $\alpha > \beta$ is true at w' if and only if $*(w', \|\alpha\|^{\mathfrak{M}}) \subseteq \|\beta\|^{\mathfrak{M}}$. What matters is not that it is a particular possible world that validates $\alpha > \beta$. Rather, it is the

[1]We adopt here Asher and Morreau's symbolism of selection function.

"meaning" of $\alpha > \beta$, which can be expressed as a proposition $\|\alpha > \beta\|^{\mathfrak{M}}$ (i.e., a set of possible worlds where $\alpha > \beta$ is true). It inspires the idea that the first argument of $*$ might be lifted to a proposition as well. Denoting the new selection function as $\circledast$, lifting $*$ to $\circledast$ makes the latter to be a function from $\wp(W) \times \wp(W)$ to $\wp(W)$. To differentiate the new selection function from what is used in the literature by Nute, Stalnaker, Delgrande, Asher, Morreau, etc. (i.e., $*$: $W \times \wp(W) \to \wp(W)$), let us call the new one the *set* selection function and theirs the *point* selection function.

$\circledast(\|\alpha > \beta\|^{\mathfrak{M}}, \|\alpha\|^{\mathfrak{M}}) \subseteq \|\beta\|^{\mathfrak{M}}$ suggests a reading that in all worlds where $\alpha > \beta$ "applies" to α, β is true. In other words, applying default rule $\alpha > \beta$ onto its antecedent α, we can get its consequent β. This fits our intuitive meaning of $\alpha > \beta$ very well. Unfortunately, we cannot use it as a truth condition; e.g., $\alpha > \beta$ is true at w iff $w \in \|\alpha > \beta\|^{\mathfrak{M}}$ where $\circledast(\|\alpha > \beta\|^{\mathfrak{M}}, \|\alpha\|^{\mathfrak{M}}) \subseteq \|\beta\|^{\mathfrak{M}}$. Otherwise, we will find ourselves in a circular situation in which the truth value of $\alpha > \beta$ depends on the value of $\circledast$ function, which in fact takes the overall truth value of $\alpha > \beta$ in all of the possible worlds as input. The best we can do to avoid this circularity is to make $\circledast(\|\alpha > \beta\|^{\mathfrak{M}}, \|\alpha\|^{\mathfrak{M}}) \subseteq \|\beta\|^{\mathfrak{M}}$ the provable property naturally falling out of the truth condition for $\alpha > \beta$ via the $\circledast$ function. We propose that

$$\|\alpha > \beta\|^{\mathfrak{M}} = \bigcup\{X \subseteq W \mid \circledast(X, \|\alpha\|^{\mathfrak{M}}) \subseteq \|\beta\|^{\mathfrak{M}}\},$$

imaging that every X satisfying $\circledast(X, \|\alpha\|^{\mathfrak{M}}) \subseteq \|\beta\|^{\mathfrak{M}}$ is an "area" where $\alpha > \beta$ should have been committed. Intuitively, $\|\alpha > \beta\|^{\mathfrak{M}}$ "collects" all "areas" where $\alpha > \beta$ is true, and sets up the largest boundary possible for $\alpha > \beta$ being true. In order to prove $\circledast(\|\alpha > \beta\|^{\mathfrak{M}}, \|\alpha\|^{\mathfrak{M}}) \subseteq \|\beta\|^{\mathfrak{M}}$, we need to make sure that set $\{X \subseteq W \mid \circledast(X, \|\alpha\|^{\mathfrak{M}}) \subseteq \|\beta\|^{\mathfrak{M}}\}$ is closed under union. This will require some frame constraints on the $\circledast$ function. The details of the semantics will be worked out in the next section.

To have a better understanding of the lifted selection function $\circledast$, we may draw an analogy with the ternary relation used in Meyer and Routley's semantics for relevance logic. Meyer and Routley explain the intuition behind the use of $Rw_1w_2w_3$ in this way (cf., [4], p164)[2]: Relative to the laws in w_1, $Rw_1w_2w_3$ means that w_3 is accessible from w_2; i.e., if the antecedent of w_1-law is realized in w_2, then its consequent is realized in w_3. This is to say that if a relevance implication $\alpha \mapsto \beta$ is true at w_1 and its antecedent α is true at w_2 then its consequent β must be true at w_3. The function $\circledast$ can be converted to a ternary relation $R_{\circledast}$: $\circledast(X, Y) \subseteq Z$ iff $R_{\circledast}XYZ$. We may phrase the following reading for $R_{\circledast}XYZ$: if a default law $\alpha > \beta$ is true at all worlds in X, and its antecedent α is true at all worlds in Y, then

[2] a, b and c in their original writings are replaced with w_1, w_2 and w_3 respectively, to stand for the possible worlds.

its consequent β is true at all worlds in Z. Given any formula φ and any proposition X, if X exactly contains those possible worlds where φ is true, then we say that φ *expresses* X. A proposition X is *expressible* if there is a formula φ such that φ expresses X. If X, Y and Z, arguments in the relation $R_{\circledast}XYZ$, are seen not only literally as sets of possible worlds, but also propositions that are expressible by particular formulas, then $R_{\circledast}XYZ$ can be understood as this: if X is a proposition expressible by a default law $\alpha > \beta$ and Y expressible by α, then the proposition that β expresses must be included in Z. The combination of the information regarding a default law and its antecedent brings about the corresponding consequent. This is how we *apply* a default law to the places where its antecedent is realized. The "applying" procedure to combine a default law to its antecedent may take various forms; one of them can simply be that $\alpha > \beta$ and α are true in the same world. In that case, we can adopt the following as a frame constraint:

$$\text{For all } Z \subseteq W, \text{ if} \circledast (X,Y) \subseteq Z, \text{ then} \circledast (W, X \cap Y) \subseteq Z.$$

Since $X \cap Y$ holds the combined information regarding the default law $\alpha > \beta$ and its antecedent α, no more additional information is needed elsewhere. Thus, the first argument of $\circledast$ in the "then" clause is trivialized to W (where no specific information is held). This constraint validates $(\alpha \wedge (\alpha > \beta)) > \beta$. A detailed proof will be shown in the next section. Using the $\circledast$ function, the frame condition for $(\alpha \wedge (\alpha > \beta)) > \beta$ is quite simple. In addition, there is also a clear intuition backing up this constraint. Recall the distinction made by Delgrande between defaults (considered as background knowledge) and contingent facts (considered as evidence). The constraint blurs the distinction by allowing to merge some defaults into facts (since they are expressed by sentences just as facts are). There is a clear increase in expressive power offered by such a conflation, for it allows one to achieve a degree of interaction between the two components that is otherwise unobtainable. Besides various putative advantages of the set selection function $\circledast$ may bring to us, its main contribution is to remove the distinction between background and evidence at this logical level.

If we do not lift $*$ to $\circledast$ and just take over the original interpretation of $>$, $(\alpha \wedge (\alpha > \beta)) > \beta$ is not valid without some frame constraints. A simple frame condition like if $w \in S$ then $w \in *(w, S)$ together with facticity (i.e., $*(w, S) \subseteq S$) could validate $(\alpha \wedge (\alpha > \beta)) > \beta$[3]. However, this frame condition will make $(\alpha \wedge (\alpha >$

[3]Here is a proof for $(\alpha \wedge (\alpha > \beta)) > \beta$ being valid under the frame condition if $w \in S$ then $w \in *(w, S)$. For any w, w', suppose that $w' \in *(w, \|\alpha \wedge (\alpha > \beta)\|)$. By the facticity, $w' \in \|\alpha \wedge (\alpha > \beta)\| = \|\alpha\| \cap \|\alpha > \beta\|$. Since $w' \in \|\alpha > \beta\|$, $*(w', \|\alpha\|) \subseteq \|\beta\|$. Since $w' \in \|\alpha\|$, $w' \in *(w', \|\alpha\|) \subseteq \|\beta\|$. Therefore, $*(w, \|\alpha \wedge (\alpha > \beta)\|) \subseteq \|\beta\|$. Hence, $w \in \|(\alpha \wedge (\alpha > \beta)) > \beta\|$.

$\beta)) \wedge \neg\beta$ incompatible[4] in the meantime. It does not seem to be straightforward to find a frame condition in the original semantics that validates $(\alpha \wedge (\alpha > \beta)) > \beta$ but still makes $(\alpha \wedge (\alpha > \beta)) \wedge \neg\beta$ compatible. It is even harder to require the frame condition of the point selection semantics to match the intuition of "normally follow" as our $\circledast$-based semantics does.

In terms of set selection function, the frame condition for IDENTITY axiom $\alpha > \alpha$ is $\circledast(X, Y) \subseteq Y$. Considering $\bigcup\{Z \subseteq W \mid \circledast(Z, X) \subseteq Y\}$ (denoted by $[X, Y]$) as the proposition that states a default of which the antecedent proposition is X and the consequent proposition is Y, the frame condition for RPT can be expressed by this if-then clause: If $\circledast(X, Y \cap Y') \subseteq Z$, then $\circledast(X, Y \cap [Y, Y'] \cap [Y', Z]) \subseteq Z$. We will give formal definitions and proofs in detail in the section to come.

3.2 $\mathfrak{L}_>$-Frames and Models

Formulas of $\mathfrak{L}_>$ are interpreted in $\mathfrak{L}_>$-models, each of which consists of an $\mathfrak{L}_>$-frame and a truth value assignment function. An $\mathfrak{L}_>$-frame is an ordered pair of a set of possible worlds and a *set selection* function (i.e., $\circledast$: $\wp(W) \times \wp(W) \to \wp(W)$) defined upon it.

Definition 1. *$\circledast$ is a set selection function defined on a set of possible worlds W: $\wp(W) \times \wp(W) \to \wp(W)$.*

Definition 2. *Given any $X, Y \subseteq W$, let $[X, Y] = \bigcup\{Z \subseteq W \mid \circledast(Z, X) \subseteq Y\}$.*

Definition 3. *An $\mathfrak{L}_>$-frame is an ordered pair $\mathfrak{F} = \langle W, \circledast \rangle$ where*

1. *W is a non-empty set of possible worlds;*
2. *$\circledast$ is a set selection function defined on W satisfying:*
   (a) *If $X \subseteq X'$, then $\circledast(X, Y) \subseteq \circledast(X', Y)$;*
   (b) *If $\circledast(\{w\}, Y) \subseteq Z$ for every $w \in X$, then $\circledast(X, Y) \subseteq Z$;*
   (c) *If $\circledast(X, Y) \subseteq Z$, then $\circledast(W, X \cap Y) \subseteq Z$;*
   (d) *$\circledast(X, Y) \subseteq Y$;*
   (e) *If $\circledast(X, Y \cap Y') \subseteq Z$, then $\circledast(X, Y \cap [Y, Y'] \cap [Y', Z]) \subseteq Z$.*

[4] A proof for $(\alpha \wedge (\alpha > \beta)) \wedge \neg\beta$ being incompatible, under the frame condition if $w \in S$ then $w \in *(w, S)$, can be constructed like this: For any w, suppose $w \in \|(\alpha \wedge (\alpha > \beta))\| = \|\alpha\| \cap \|\alpha > \beta\|$. Then, since $w \in \|\alpha > \beta\|$, $*(w, \|\alpha\|) \subseteq \|\beta\|$. Since $w \in \|\alpha\|$, $w \in *(w, \|\alpha\|) \subseteq \|\beta\|$. Therefore, $\|\alpha \wedge (\alpha > \beta)\| \subseteq \|\beta\|$. Hence, $\|(\alpha \wedge (\alpha > \beta)) \wedge \neg\beta\| = \|\alpha \wedge (\alpha > \beta)\| \cap \|\neg\beta\| = \varnothing$.

Let I be a set of indexes. From the frame conditions 2(a) and 2(b) listed above, it is not difficult to see that the following condition is satisfied on an $\mathfrak{L}_>$-frame:

$$\circledast(\bigcup\{X_i : i \in I\}, Y) \subseteq \bigcup\{\circledast(X_i, Y) : i \in I\} \qquad \text{(Frame Condition (b'))}$$

The set selection function $\circledast$ employed here is closely related to the point selection function. Given a point selection function $*$ on W: $W \times \wp(W) \to \wp(W)$, we can convert it into a set selection function $\circledast_*$ as follows:

$$\circledast_*(X, Y) = \bigcup_{w \in X} *(w, Y)$$

This set selection function obtained from conversion will automatically satisfy the frame condition 2(a) and 2(b) in Definition 3.

Conversely, given a set selection function $\circledast$, we can convert it into a point selection function $*_\circledast$ as well:

$$*_\circledast(w, Y) = \circledast(\{w\}, Y)$$

The conversion from $\circledast$ to $*_\circledast$ loses some information. $*_\circledast$ is $\circledast$ restricted on singleton sets as its first argument. As a result, the flexibility provided by the first argument of $\circledast$ disappears in $*_\circledast$, and hence the counterpart of some simple condition for $\circledast$ will look quite awkward in terms of $*_\circledast$. For example, condition $\circledast(W, X \cap Y) \subseteq \circledast(X, Y)$ for $\circledast$ becomes its counterpart below for $*_\circledast$:

$$\bigcup_{w \in W} *_\circledast(\{w\}, X \cap Y) \subseteq \bigcup_{w \in X} *_\circledast(\{w\}, Y)$$

We have shown that $\circledast$ is a more general function than $*$ in the sense that the former accommodates the latter as a special case. This enhanced power facilitates spelling out the frame condition 2(c), which as we will see shortly, validates $(\alpha \wedge (\alpha > \beta)) > \beta$.

Frame conditions 2(d) and 2(e) validate the ID and RPT, respectively.

Definition 4. *An $\mathfrak{L}_>$-model is an ordered pair $\mathfrak{M} = \langle \mathfrak{F}, \sigma \rangle$ where*

1. *$\mathfrak{F}$ is an $\mathfrak{L}_>$-frame, and*
2. *σ is a truth value assignment function from P to $\wp(W)$.*

We can also write $\mathfrak{M}$ as a triple $\langle W, \circledast, \sigma \rangle$. Given $\mathfrak{M}$, we use $W_\mathfrak{M}$, $\circledast_\mathfrak{M}$ and $\sigma_\mathfrak{M}$ to denote three components of $\mathfrak{M}$, respectively.

Definition 5. *For any formula α, the symbolism $\|\alpha\|^{\mathfrak{M}}$ is used to stand for the set of worlds in $\mathfrak{M}$ in which α is true, satisfying:*

1. $\|p\|^{\mathfrak{M}} = \sigma(p)$
2. $\|\neg\alpha\|^{\mathfrak{M}} = W - \|\alpha\|^{\mathfrak{M}}$
3. $\|\alpha \to \beta\|^{\mathfrak{M}} = (W - \|\alpha\|^{\mathfrak{M}}) \cup \|\beta\|^{\mathfrak{M}}$
4. $\|\alpha > \beta\|^{\mathfrak{M}} = \bigcup\{X \subseteq W \mid \circledast(X, \|\alpha\|^{\mathfrak{M}}) \subseteq \|\beta\|^{\mathfrak{M}}\}$

Items 1-3 in the above definition are straightforward. Item 4 needs some explanation. The Frame Condition (b'), which is provable from Definition 3-2(a) and 2(b), guarantees that $\circledast(\|\alpha > \beta\|^{\mathfrak{M}}, \|\alpha\|^{\mathfrak{M}}) \subseteq \bigcup\{\circledast(X, \|\alpha\|^{\mathfrak{M}}) \mid \circledast(X, \|\alpha\|^{\mathfrak{M}}) \subseteq \|\beta\|^{\mathfrak{M}}\} \subseteq \|\beta\|^{\mathfrak{M}}$. The equation in item 4 actually says that $\|\alpha > \beta\|^{\mathfrak{M}}$ is the largest set X such that $\circledast(X, \|\alpha\|^{\mathfrak{M}}) \subseteq \|\beta\|^{\mathfrak{M}}$. Any set X' satisfying $\circledast(X', \|\alpha\|^{\mathfrak{M}}) \subseteq \|\beta\|^{\mathfrak{M}}$ should be a subset of $\|\alpha > \beta\|^{\mathfrak{M}}$; that is, $\alpha > \beta$ is true on X'. The intuition behind this is that if β is true in all selected possible worlds relative to the proposition α by function $\circledast$ from the point of view of X', then $\alpha > \beta$ is a default that holds on X'. On the other hand, for any $X' \subseteq \|\alpha > \beta\|^{\mathfrak{M}}$, we have $\circledast(X', \|\alpha\|^{\mathfrak{M}}) \subseteq \|\beta\|^{\mathfrak{M}}$ due to the first property of $\circledast$ (see Definition 3-2(a)). This is to say: if $\alpha > \beta$ is a default that holds on X, then "applying" the default $\alpha > \beta$ onto α will get us back the conclusion β. With $\|\alpha > \beta\|^{\mathfrak{M}}$ so defined, it has two important properties:

$$\circledast(\|\alpha > \beta\|^{\mathfrak{M}}, \|\alpha\|^{\mathfrak{M}}) \subseteq \|\beta\|^{\mathfrak{M}} \qquad (\|\alpha \geqslant \beta\|^{\mathfrak{M}} \text{ property 1})$$
$$X \subseteq \|\alpha > \beta\|^{\mathfrak{M}} \text{ iff } \circledast(X, \|\alpha\|^{\mathfrak{M}}) \subseteq \|\beta\|^{\mathfrak{M}} \qquad (\|\alpha \geqslant \beta\|^{\mathfrak{M}} \text{ property 2})$$

This may be a good place to explain and highlight the advantage of using a set selection function instead of a point selection function. The first parameter of the set selection function is intended to capture the set of possible worlds where $\alpha > \beta$ is true. It makes sense to take the intersection of two arguments X and Y of function $\circledast$, which are now two sets. Thinking of X and Y in terms of propositions, if X and Y are places where $\alpha > \beta$ and α are true respectively, then $X \cap Y$ represents the place where $(\alpha > \beta) \wedge \alpha$ is true. This is not possible when X is not a set but a single possible world.

The set selection function makes it easy to express the third frame condition $\circledast(W, X \cap Y) \subseteq \circledast(X, Y)$ (see Definition 3-2(c)). $\circledast(W, X \cap Y) \subseteq \circledast(X, Y)$ is equivalent to, for all Z, if $\circledast(X, Y) \subseteq Z$ then $\circledast(W, X \cap Y) \subseteq Z$. When X, Y and Z are propositions of $\alpha > \beta$, α and β (i.e., $X = \|\alpha > \beta\|$, $Y = \|\alpha\|$ and $Z = \|\beta\|$) in particular, this condition says that $W \subseteq \|((\alpha > \beta) \wedge \alpha) > \beta\|$. This means that

$((\alpha > \beta) \wedge \alpha) > \beta$ is true everywhere. It is not surprising that the third frame condition validates $((\alpha > \beta) \wedge \alpha) > \beta$. Using a point selection function whose first argument is a single possible world, this idea cannot be spelled out so easily.

Conceptually, we are convinced that the axiom DMP should be introduced to investigate the type of "normally follow" inferences. Technically, we have improved the traditional selection function and do not encounter any more the problem of having awkward semantic characterization for DMP. In fact, the frame condition of DMP using set selection function appears quite natural, and it supports, from the semantic point of view, our choice of DMP as an axiom. We will complete the presentation of a formal semantics in the next section and then prove some meta properties of the system DC (soundness and completeness) to show that the formal semantics characterizes the system DC.

3.3 Satisfiability and Validity

The symbolism $\mathfrak{M} \models_X \alpha$ is introduced to assert that α *is true at the set of worlds* X *in the model* $\mathfrak{M}$.

Definition 6. *Let* $\mathfrak{M}$ *be an* $\mathfrak{L}_{>}$*-model,* X *be a non-empty subset of* $W_{\mathfrak{M}}$ *and* α *be a formula.* $\mathfrak{M} \models_X \alpha$ *iff* $X \subseteq \|\alpha\|^{\mathfrak{M}}$.

Proposition 1. *The following holds due to the definition of* $\mathfrak{M} \models_X \alpha$*, where* X *in 3, 4 is* $\{w\}$ *for some* $w \in W_{\mathfrak{M}}$ *and in 1, 2, 5 is any non-empty subset of* $W_{\mathfrak{M}}$.

1. $\mathfrak{M} \models_X \neg\alpha$ *iff* $X \cap \|\alpha\|^{\mathfrak{M}} = \varnothing$.
2. $\mathfrak{M} \models_X \alpha \wedge \beta$ *iff* $\mathfrak{M} \models_X \alpha$ *and* $\mathfrak{M} \models_X \beta$.
3. $\mathfrak{M} \models_X \alpha \vee \beta$ *iff* $\mathfrak{M} \models_X \alpha$ *or* $\mathfrak{M} \models_X \beta$.
4. $\mathfrak{M} \models_X \alpha \to \beta$ *iff if* $\mathfrak{M} \models_X \alpha$ *then* $\mathfrak{M} \models_X \beta$.
5. $\mathfrak{M} \models_X \alpha > \beta$ *iff* $\circledast(X, \|\alpha\|^{\mathfrak{M}}) \subseteq \|\beta\|^{\mathfrak{M}}$.

Here is a proof for item 5. The rest of the items are left to the reader.

Proof. ("if" part): Suppose that $\mathfrak{M} \models_X \alpha > \beta$. By Definition 6, $X \subseteq \|\alpha > \beta\|^{\mathfrak{M}}$. By $\|\alpha > \beta\|^{\mathfrak{M}}$ property 2, $\circledast(X, \|\alpha\|^{\mathfrak{M}}) \subseteq \|\beta\|^{\mathfrak{M}}$.

("only if" part): Suppose that $\circledast(X, \|\alpha\|^{\mathfrak{M}}) \subseteq \|\beta\|^{\mathfrak{M}}$. By Definition 5(4), $X \subseteq \|\alpha > \beta\|^{\mathfrak{M}}$. By Definition 6, $\mathfrak{M} \models_X \alpha > \beta$. □

Proposition 2. *For each* $X \subseteq W_{\mathfrak{M}}$, $\mathfrak{M} \models_X \alpha \to \beta$ *iff* $X \cap \|\alpha\|^{\mathfrak{M}} \subseteq \|\beta\|^{\mathfrak{M}}$. *In particular,* $\mathfrak{M} \models_{W_{\mathfrak{M}}} \alpha \to \beta$ *iff* $\|\alpha\|^{\mathfrak{M}} \subseteq \|\beta\|^{\mathfrak{M}}$.

Proposition 3. *$\mathfrak{M} \models_{W_\mathfrak{M}} \alpha$ iff $\mathfrak{M} \models_{\{w\}} \alpha$ for all $w \in W_\mathfrak{M}$.*

Definition 7. *(Validity) Given a formula α, we say that α is* valid *(written as $\models \alpha$) if and only if α is true at every world in all models (i.e., $\mathfrak{M} \models_{W_\mathfrak{M}} \alpha$, for all $\mathfrak{M}$).*

Definition 8. *(Satisfiability) Given a formula α, α is* satisfiable *if and only if $\neg\alpha$ is not valid.*

The theorem below shows the validity of some $>$-formulas.

Theorem 2. *The following statements hold:*

1. $(\alpha > (\beta \rightarrow \gamma)) \rightarrow ((\alpha > \beta) \rightarrow (\alpha > \gamma))$ *is valid;*
2. $(\alpha \wedge (\alpha > \beta)) > \beta$ *is valid;*
3. $\alpha > \alpha$ *is valid;*
4. $((\alpha \wedge \beta) > \gamma) \rightarrow ((\alpha \wedge (\alpha > \beta) \wedge (\beta > \gamma)) > \gamma)$ *is valid.*

Neither $(\alpha > \beta) \rightarrow (\alpha \rightarrow \beta)$ nor $(\alpha \rightarrow \beta) \rightarrow (\alpha > \beta)$ is valid. This is because the reference world where $\alpha \rightarrow \beta$ is evaluated does not have to be one of the selected (normal) worlds, let alone the only normal world to be selected. If the current world had to be one of the normal worlds, then the defaults would not tolerate any exceptions. If the current world were the only normal world to be selected, then the notion of normality would collapse and the intensionality of defaults would vaporize. Notice that the requirement of validating $(\alpha \rightarrow \beta) \rightarrow (\alpha > \beta)$ is much stronger than that of preserving the validity from $\alpha \rightarrow \beta$ to $\alpha > \beta$. The latter is supposed to be the job of the derived rule RIN. The soundness of the system that will be proved shortly guarantees that RIN does preserve validity.

4 Soundness and Completeness of DC

The system DC is sound and complete with respect to the semantics laid out in Section 3.2. The results are stated in the next two theorems.

Theorem 3. *(Soundness) Given any formula α, if $\vdash_{DC} \alpha$ then $\models \alpha$.*

The proof for soundness is routine: each of axiom schemata is valid under the semantics and each of the inference rules preserves validity. The former is stated in Theorem 2. It is easy to show that the rules of inference (i.e., MP, REQ, RN, RM) preserve validity. The proof is omitted.

Theorem 4. *(Completeness) Given any formula α, if $\nvdash_{DC} \alpha$ then $\nvDash \alpha$.*

Completeness of DC is proved by constructing a canonical structure $\langle W_{DC}, \circledast_{DC}\rangle$ and a small (finite) model $\langle W_\Gamma, \circledast_\Gamma, \sigma_\Gamma\rangle$ with $\Gamma = \{\neg\alpha\}$, where α is a given formula not provable in DC. Then $\neg\alpha$ is true at some world in W_Γ.

We define $\circledast_{DC}$ merely on those expressible subsets of possible worlds W_{DC}. $\langle W_{DC}, \circledast_{DC}\rangle$ is called a *canonical structure*, but not yet a canonical frame. We will show that all of the frame conditions required in Definition 3 are satisfied in such a structure for all expressible subsets of W_{DC}. We then further construct a small model by coercing the big canonical structure into a finite frame via a homomorphism so that each subset of possible worlds in the small model is expressible. The requested frame conditions hold on the small model. The construction of the small model depends on the given unprovable formula α. When α varies, the small model to falsify it changes accordingly.

The decidability of the system DC comes out as a side product of this completeness proof. Since we can construct a finite model to satisfy any unprovable formula α of the system DC, the system DC is decidable.

In what follows, we will sketch the main definitions, lemmas, propositions and intermediate theorems that lead to the proof of completeness theorem. $|\alpha|_{DC}$ will be used to stand for the class of maximally-consistent sets of DC containing α.

Definition 9. *The canonical structure $\mathfrak{S}_{DC} = \langle W_{DC}, \circledast_{DC}\rangle$ of DC is defined as follows:*

1. $W_{DC} = \{w \mid w$ *is a DC maximally consistent set of formulas*$\}$,
2. $\circledast_{DC}(X, |\alpha|_{DC}) = \bigcap\{|\beta|_{DC} \mid X \subseteq |\alpha > \beta|_{DC}\}$ *where* $|\alpha|_{DC} = \{w \mid (w \in W_{DC}) \wedge (\alpha \in w)\}$.

Note that $\circledast_{DC}$ is not a total function defined on $\wp(W_{DC}) \times \wp(W_{DC})$. $\mathfrak{S}_{DC}$ is not a frame. We refer it as canonical structure.

Lemma 1. *If Γ is a consistent set of formulas of $\mathfrak{L}_>$ that contains the formula $\alpha \not> \beta$, then $\{\gamma \mid \alpha > \gamma \in \Gamma\} \cup \{\neg\beta\}$ is consistent.*

Proof. Let $\Lambda = \{\gamma \mid \alpha > \gamma \in \Gamma\}$ and assume that $\Lambda \cup \{\neg\beta\}$ is not consistent. So there are $\beta_1, \dots, \beta_n$ of Λ such that $\vdash \neg(\beta_1 \wedge \dots \wedge \beta_n \wedge \neg\beta)$. So, by the classical propositional logic, $\vdash \beta_1 \wedge \dots \wedge \beta_n \rightarrow \beta$. By RCK, we have $\vdash \alpha > \beta_1 \wedge \dots \wedge \alpha > \beta_n \rightarrow \alpha > \beta$. So, $\vdash \neg(\alpha > \beta_1 \wedge \dots \wedge \alpha > \beta_n \wedge (\alpha \not> \beta))$. This means that $\{\alpha > \beta_1, \dots, \alpha > \beta_n, \alpha \not> \beta\}$ is not consistent, which contradicts the premise that $\Lambda \cup \{(\alpha \not> \beta)\} \subseteq \Gamma$ is consistent. $\square$

Theorem 5. *Let $\mathfrak{S}_{DC} = \langle W_{DC}, \circledast_{DC}\rangle$ be the canonical structure of DC. Then for each $\alpha > \beta \in \mathfrak{L}_{>}$ and each $X \subseteq W_{DC}$, $\circledast_{DC}(X, |\alpha|_{DC}) \subseteq |\beta|_{DC}$ iff $X \subseteq |\alpha > \beta|_{DC}$.*

Proof. Assume that $X \subseteq |\alpha > \beta|_{DC}$. $\circledast_{DC}(X, |\alpha|_{DC}) \subseteq |\beta|_{DC}$, by the definition of $\circledast_{DC}$.

Conversely suppose that $X \nsubseteq |\alpha > \beta|_{DC}$. Then there is $w \in X$ and $w \notin |\alpha > \beta|_{DC}$. Since w is maximally consistent, $(\alpha \not> \beta) \in w$. By Lemma 1, $\{\gamma \mid \alpha > \gamma \in w\} \cup \{\neg\beta\}$ is consistent. Since $w \in X$, $\{\gamma \mid X \subseteq |\alpha > \gamma|_{DC}\} \subseteq \{\gamma \mid w \in |\alpha > \gamma|_{DC}\} = \{\gamma \mid \alpha > \gamma \in w\}$. Thus, $\{\gamma \mid X \subseteq |\alpha > \gamma|_{DC}\} \cup \{\neg\beta\}$ is consistent as well. By the Lindenbaum's lemma, it has a maximally consistent extension w_1 such that $\{\gamma \mid X \subseteq |\alpha > \gamma|_{DC}\} \subseteq w_1$ and $\neg\beta \in w_1$. By the definition of $\circledast_{DC}$, $w_1 \in \circledast_{DC}(X, |\alpha|_{DC})$. Nevertheless, since $\neg\beta \in w_1$, $w_1 \notin |\beta|_{DC}$. Thus, $\circledast_{DC}(X, |\alpha|_{DC}) \nsubseteq |\beta|_{DC}$. □

Theorem 6. $[|\alpha|_{DC}, |\beta|_{DC}]_{DC} = |\alpha > \beta|_{DC}$

Proof. According to Definition 2, $[|\alpha|_{DC}, |\beta|_{DC}]_{DC} = \bigcup\{X \subseteq W_{DC} \mid \circledast_{DC}(X, |\alpha|_{DC}) \subseteq |\beta|_{DC}\}$. By Theorem 5, for any $X \subseteq W_{DC}$, if $X \subseteq |\alpha > \beta|_{DC}$, then $\circledast_{DC}(X, |\alpha|_{DC}) \subseteq |\beta|_{DC}$. Hence, $X \subseteq [|\alpha|_{DC}, |\beta|_{DC}]_{DC}$. Therefore, $|\alpha > \beta|_{DC} \subseteq [|\alpha|_{DC}, |\beta|_{DC}]_{DC}$.

To show the other direction of the inclusion, let us take any arbitrary $w \in W_{DC}$ such that $w \in [|\alpha|_{DC}, |\beta|_{DC}]_{DC}$. There must be an $X \subseteq W_{DC}$ such that $w \in X$ and $\circledast_{DC}(X, |\alpha|_{DC}) \subseteq |\beta|_{DC}$. By Theorem 5, $X \subseteq |\alpha > \beta|_{DC}$. So, $w \in |\alpha > \beta|_{DC}$. □

Theorem 7. *The canonical structure $\mathfrak{S}_{DC} = \langle W_{DC}, \circledast_{DC}\rangle$ satisfies:*

1. *If $X \subseteq X' \subseteq W_{DC}$, then $\circledast_{DC}(X, |\alpha|_{DC}) \subseteq \circledast_{DC}(X', |\alpha|_{DC})$;*
2. *If $\circledast(\{w\}, |\alpha|_{DC}) \subseteq |\beta|_{DC}$ for every $w \in X \subseteq W_{DC}$, then $\circledast(X, |\alpha|_{DC}) \subseteq |\beta|_{DC}$;*
3. *If $\circledast_{DC}(|\gamma|_{DC}, |\alpha|_{DC}) \subseteq |\beta|_{DC}$ then $\circledast_{DC}(W_{DC}, |\gamma|_{DC} \cap |\alpha|_{DC}) \subseteq |\beta|_{DC}$;*
4. *$\circledast_{DC}(X, |\alpha|_{DC}) \subseteq |\alpha|_{DC}$;*
5. *If $\circledast_{DC}(X, |\alpha|_{DC} \cap |\beta|_{DC}) \subseteq |\gamma|_{DC}$, then $\circledast_{DC}(X, |\alpha|_{DC} \cap |\alpha > \beta|_{DC} \cap |\beta > \gamma|_{DC}) \subseteq |\gamma|_{DC}$.*

Proof. 1: Suppose that $X \subseteq X'$, and $w \in \circledast_{DC}(X, |\alpha|_{DC})$. Let $X' \subseteq |\alpha > \beta|_{DC}$, where β is an arbitrary formula. Since $X \subseteq X'$, $X \subseteq |\alpha > \beta|_{DC}$. Since $w \in \circledast_{DC}(X, |\alpha|_{DC})$, by Definition 9(2), $\beta \in w$. Thus, $\{\beta \mid X' \subseteq |\alpha > \beta|_{DC}\} \subseteq w$. By Definition 9(2), $w \in \circledast_{DC}(X', |\alpha|_{DC})$.

2: Suppose that $\circledast(\{w\}, |\alpha|_{DC}) \subseteq |\beta|_{DC}$ for every $w \in X \subseteq W_{DC}$ and that $\circledast(X, |\alpha|_{DC}) \nsubseteq |\beta|_{DC}$. There must be some $w_0 \in X$ such that $w_0 \notin |\alpha > \beta|_{DC}$ (that is, $\{w_0\} \nsubseteq |\alpha > \beta|_{DC}$). By Theorem 5, $\circledast(\{w_0\}, |\alpha|_{DC}) \nsubseteq |\beta|_{DC}$, which contradicts our assumption.

3: Suppose that $\circledast_{DC}(|\gamma|_{DC}, |\alpha|_{DC}) \subseteq |\beta|_{DC}$ and that $w \in \circledast_{DC}(W_{DC}, |\gamma|_{DC} \cap |\alpha|_{DC})$. By Theorem 5, $|\gamma|_{DC} \subseteq |\alpha > \beta|_{DC}$. Then $|\gamma|_{DC} = |\gamma|_{DC} \cap |\alpha > \beta|_{DC}$. Then $|\gamma \wedge \alpha|_{DC} = |\gamma|_{DC} \cap |\alpha|_{DC} = |\gamma|_{DC} \cap |\alpha > \beta|_{DC} \cap |\alpha|_{DC} = |\alpha \wedge (\alpha > \beta) \wedge \gamma|_{DC}$. Since $(\alpha \wedge (\alpha > \beta)) > \beta$ is an axiom of DC, by the rule RM, $(\alpha \wedge (\alpha > \beta) \wedge \gamma) > \beta$ is a theorem of DC. Thus, $W_{DC} \subseteq |((\alpha \wedge (\alpha > \beta) \wedge \gamma)) > \beta|_{DC} = |(\gamma \wedge \alpha) > \beta|_{DC}$. By Definition 9(2), $\beta \in w$ since $w \in \circledast_{DC}(W_{DC}, |\gamma|_{DC} \cap |\alpha|_{DC})$.

4: Suppose that $w \in \circledast_{DC}(X, |\alpha|_{DC})$. Since $\alpha > \alpha$ is an axiom of DC, $X \subseteq |\alpha > \alpha|_{DC} = W_{DC}$. Thus, $\alpha \in w$, and hence $w \in |\alpha|_{DC}$.

5: $|(\alpha \wedge \beta) > \gamma|_{DC} \subseteq |(\alpha \wedge (\alpha > \beta) \wedge (\beta > \gamma)) > \gamma|_{DC}$, due to the axiom RPT. Thus, for any $X \subseteq W_{DC}$, if $X \subseteq |(\alpha \wedge \beta) > \gamma|_{DC}$, we have $X \subseteq |(\alpha \wedge (\alpha > \beta) \wedge (\beta > \gamma)) > \gamma|_{DC}$. By Theorem 5, this is to say, if $\circledast_{DC}(X, |\alpha|_{DC} \cap |\beta|_{DC}) \subseteq |\gamma|_{DC}$ then $\circledast_{DC}(X, |\alpha|_{DC} \cap |\alpha > \beta|_{DC} \cap |\beta > \gamma|_{DC}) \subseteq |\gamma|_{DC}$. □

This theorem states that all frame conditions required in Definition 3 are satisfied in the canonical structure for those subsets of W_{DC} that are expressible by formulas. This is a good result, but it is not strong enough to justify the completeness theorem that we seek. Next, we will construct a finite model based on the canonical structure and preserve this good result to the finite model where all subsets of its possible worlds are expressible.

Definition 10. *(closed under single negation) Given a formula α, $\sim\alpha$ is defined as the following formula:*

$$\sim\alpha = \begin{cases} \beta \ , \text{ if } \alpha \text{ is of the form } \neg\beta, \\ \neg\alpha, \text{ otherwise} \end{cases}$$

A set Γ of formulas is closed under single negation if $\sim\alpha$ belongs to Γ whenever $\alpha \in \Gamma$.

Definition 11. *(component closure) Given a set Γ of formulas, the component closure of Γ, written as $\mathsf{CCL}(\Gamma)$, is the smallest set of formulas containing Γ that is closed under subformulas and single negation.*

It is crucial to note that if Γ is finite, then so is $\mathsf{CCL}(\Gamma)$.

Definition 12. *(restricted MCS) Let Γ be a set of formulas. A restricted maximally consistent set A over Γ is*

1. *a subset of* $\mathsf{CCL}(\Gamma)$*;*

2. *a consistent set;*

3. *if* $A \subseteq B \subseteq \mathsf{CCL}(\Gamma)$ *then* $A = B$.

Let W_Γ denote the set containing all restricted maximally consistent sets over Γ. When Γ is finite, so are W_Γ and each element of W_Γ. It is worth noting that for any $A \in W_\Gamma$ and $\alpha \in \mathsf{CCL}(\Gamma)$, $\alpha \in A$ or $\sim \alpha \in A$.

Definition 13. *Given any* $\alpha \in \mathsf{CCL}(\Gamma)$*, let* $|\alpha|_\Gamma$ *denote all restricted maximally consistent sets over* Γ *that contain* α*. That is,* $|\alpha|_\Gamma = \{A \in W_\Gamma \mid \alpha \in A\}$.

An analogy of Lindenbaum's Lemma holds:

Proposition 4. *Given any* $\alpha \in \mathsf{CCL}(\Gamma)$*, if* α *is consistent then there exists some* $A \in W_\Gamma$ *such that* $\alpha \in A$.

Proposition 5. *For each* $\alpha \in \mathsf{CCL}(\Gamma)$ *and each* $\alpha \to \beta \in \mathsf{CCL}(\Gamma)$*, we have*

1. $|\neg\alpha|_\Gamma = W_\Gamma - |\alpha|_\Gamma$

2. $|\alpha \to \beta|_\Gamma = (W_\Gamma - |\alpha|_\Gamma) \cup |\beta|_\Gamma$

Definition 14. *Let* Γ *be a set of formulas. Given any* $A \in W_\Gamma$*, we define that* $\varphi_A = \alpha_1 \wedge \alpha_2 \wedge \ldots \wedge \alpha_n$*, where* $A = \{\alpha_1, \alpha_2, \ldots, \alpha_n\}$.

Definition 15. *Let* Γ *be a set of formulas. Given any* $X \subseteq W_\Gamma$*, we define that* $\varphi_X = \varphi_{A_1} \vee \varphi_{A_2} \vee \ldots \vee \varphi_{A_m}$*, where* $X = \{A_1, A_2, \ldots, A_m\}$.

Proposition 6. *According to Definition 15 above, we have:*

1. *for any* $X, Y \subseteq W_\Gamma$*, if* $X \subseteq Y$ *then* $|\varphi_X|_{DC} \subseteq |\varphi_Y|_{DC}$*;*

2. *for each* $\alpha \in \mathsf{CCL}(\Gamma)$*,* $\left|\varphi_{|\alpha|_\Gamma}\right|_{DC} = |\alpha|_{DC}$.

Definition 16. *Let* Γ *be a set of formulas. Define a mapping* f*:* $W_{DC} \longrightarrow W_\Gamma$*, satisfying, for each* $w_{DC} \in W_{DC}$*,* $f(w_{DC}) = w_\Gamma$ *if* $w_\Gamma \subseteq w_{DC}$.

Definition 17. *Let* g *be a mapping* $\wp(W_{DC}) \longrightarrow \wp(W_\Gamma)$ *such that, for any* $X_{DC} \in \wp(W_{DC})$*,* $g(X_{DC}) = \{f(w_{DC}) \mid w_{DC} \in X_{DC}\}$.

Definition 18. *: Let* g^* *be a mapping* $\wp(W_\Gamma) \longrightarrow \wp(W_{DC})$*, such that, for any* $X_\Gamma \in \wp(W_\Gamma)$*,* $g^*(X_\Gamma) = \{w_{DC} \in W_{DC} \mid \exists x_\Gamma \in X_\Gamma(f(w_{DC}) = x_\Gamma)\}$.

It is clear that f, g, g^* are functions. As a matter of fact, f and g are surjections, and g^* is an injection.

Proposition 7. *Here are some properties of functions g and g^*:*

1. For any X_{DC}, $X'_{DC} \subseteq W_{DC}$, if $X_{DC} \subseteq X'_{DC}$ then $g(X_{DC}) \subseteq g(X'_{DC})$

2a. For any X_Γ, $X'_\Gamma \subseteq W_\Gamma$, if $X_\Gamma \subseteq X'_\Gamma$ then $g^*(X_\Gamma) \subseteq g^*(X'_\Gamma)$

3a. For each $X_\Gamma \subseteq W_\Gamma$, $g(g^*(X_\Gamma)) \subseteq X_\Gamma$

2b. For any X_Γ, $X'_\Gamma \subseteq W_\Gamma$, if $g^*(X_\Gamma) \subseteq g^*(X'_\Gamma)$ then $X_\Gamma \subseteq X'_\Gamma$

3b. For each $X_\Gamma \subseteq W_\Gamma$, $X_\Gamma \subseteq g(g^*(X_\Gamma))$

2. For any X_Γ, $X'_\Gamma \subseteq W_\Gamma$, $X_\Gamma \subseteq X'_\Gamma$ iff $g^*(X_\Gamma) \subseteq g^*(X'_\Gamma)$

3. For each $X_\Gamma \subseteq W_\Gamma$, $g(\ g^*(X_\Gamma)) = X_\Gamma$

4. For each $X_{DC} \subseteq W_{DC}$, $X_{DC} \subseteq g^*(\ g(X_{DC}))$

5. For any X_Γ, $X'_\Gamma \subseteq W_\Gamma$, $g^*(X_\Gamma \cap X'_\Gamma) = g^*(X_\Gamma) \cap g^*(X'_\Gamma)$

6. For each $X_\Gamma \subseteq W_\Gamma$, $g^*(X_\Gamma) = \left|\varphi_{X_\Gamma}\right|_{DC}$

7. For each $X_\Gamma \subseteq W_\Gamma$, $g(|\varphi_{X_\Gamma}|_{DC}) = g(g^*(X_\Gamma)) = X_\Gamma$

8. $W_{DC} = g^*(W_\Gamma) = |\varphi_{W_\Gamma}|_{DC}$

9. For any X_Γ, $X'_\Gamma \subseteq W_\Gamma$, $|\varphi_{X_\Gamma}|_{DC} \cap |\varphi_{Y_\Gamma}|_{DC} = |\varphi_{X_\Gamma \cap Y_\Gamma}|_{DC}$

10. For each $\alpha \in \mathsf{CCL}(\Gamma)$, $g^*(|\alpha|_\Gamma) = \left|\varphi_{|\alpha|_\Gamma}\right|_{DC} = |\alpha|_{DC}$

11. For each $\alpha \in \mathsf{CCL}(\Gamma)$, $g(|\alpha|_{DC}) = |\alpha|_\Gamma$

12. For each $\alpha \in \mathsf{CCL}(\Gamma)$, $g^*(g(|\alpha|_{DC})) = |\alpha|_{DC}$

Proof. 1: Suppose that $w_\Gamma \in g(X_{DC})$ for an arbitrary $w_\Gamma \in W_\Gamma$. By the definition of g, there must be some $w_{DC} \in X_{DC}$ such that $f(w_{DC}) = w_\Gamma$. Since $X_{DC} \subseteq X'_{DC}$, $w_{DC} \in X'_{DC}$. By the definition of g again, $w_\Gamma \in g(X'_{DC})$ and hence $g(X_{DC}) \subseteq g(X'_{DC})$.

2a: Suppose that $X_\Gamma \subseteq X'_\Gamma$ and $w_{DC} \in g^*(X_\Gamma)$ for an arbitrary $w_{DC} \in W_{DC}$. By the definition of g^*, there must be some $w_\Gamma \in X_\Gamma$ such that $f(w_{DC}) = w_\Gamma$.

Since $X_\Gamma \subseteq X'_\Gamma$, $w_\Gamma \in X'_\Gamma$. By the definition of g^* again, $w_{DC} \in g^*(X'_\Gamma)$ and hence $g^*(X_\Gamma) \subseteq g^*(X'_\Gamma)$.

3a: Suppose that $w_\Gamma \in g(\ g^*(X_\Gamma))$ for an arbitrary $w_\Gamma \in W_\Gamma$. By the definition of g, there must be some $w_{DC} \in g^*(X_\Gamma)$ such that $f(w_{DC}) = w_\Gamma$. Then, by the definition of g^*, $w_\Gamma \in X_\Gamma$. So, $g(\ g^*(X_\Gamma)) \subseteq X_\Gamma$.

2b: Suppose that $g^*(X_\Gamma) \subseteq g^*(X'_\Gamma)$ and $w_\Gamma \in X_\Gamma$ for an arbitrary $w_\Gamma \in W_\Gamma$. Since $\{w_\Gamma\} \subseteq X_\Gamma$, according to 2a, $g^*(\{w_\Gamma\}) \subseteq g^*(X_\Gamma)$. Thus, $g^*(\{w_\Gamma\}) \subseteq g^*(X'_\Gamma)$. By the item 1, $g(g^*(\{w_\Gamma\})) \subseteq g(g^*(X'_\Gamma))$. By the definition of g^*, $g^*(\{w_\Gamma\}) = \{w_{DC} \in W_{DC} \mid f(w_{DC}) = w_\Gamma\}$. Then, $g(g^*(\{w_\Gamma\})) = \{w_\Gamma\}$. In addition, $g(g^*(X'_\Gamma)) \subseteq X'_\Gamma$, by 3a. So, $\{w_\Gamma\} \subseteq X'_\Gamma$ and hence $w_\Gamma \in X'_\Gamma$.

3b: Suppose that $w_\Gamma \in X_\Gamma$. By the definition of g^*, $g^*(\{w_\Gamma\}) = \{w_{DC} \in W_{DC} \mid f(w_{DC}) = w_\Gamma\}$. Thus, $g(g^*(\{w_\Gamma\})) = \{w_\Gamma\}$. Since $\{w_\Gamma\} \subseteq X_\Gamma$, according to the item 1 and 2a, $g(g^*(\{w_\Gamma\})) \subseteq g(g^*(X_\Gamma))$. So, $\{w_\Gamma\} \subseteq g(g^*(X_\Gamma))$ and hence $w_\Gamma \in g(g^*(X_\Gamma))$. Thus, $X_\Gamma \subseteq g(g^*(X_\Gamma))$.

2 can be obtained from 2a and 2b.

3 can be obtained from 3a and 3b.

4: Suppose that $w_{DC} \in X_{DC}$. Then, $f(w_{DC}) \in g(X_{DC})$. By the definition of g^*, $g^*(\{f(w_{DC})\}) = \{w'_{DC} \in W_{DC} \mid f(w'_{DC}) = f(w_{DC})\}$. Thus,$w_{DC} \in g^*(\{f(w_{DC})\})$. Since $\{f(w_{DC})\} \subseteq g(X_{DC})$, according to item 2, $g^*(\{f(w_{DC})\}) \subseteq g^*(g(X_{DC}))$. So, $w_{DC} \subseteq g^*(g(X_{DC}))$ and hence $X_{DC} \subseteq g^*(\ g(X_{DC}))$.

5: Since $(X_\Gamma \cap X'_\Gamma) \subseteq X_\Gamma$ and $(X_\Gamma \cap X'_\Gamma) \subseteq X'_\Gamma$, according to item 2, $g^*(X_\Gamma \cap X'_\Gamma) \subseteq g^*(X_\Gamma)$ and $g^*(X_\Gamma \cap X'_\Gamma) \subseteq g^*(X'_\Gamma)$. Then, $g^*(X_\Gamma \cap X'_\Gamma) \subseteq g^*(X_\Gamma) \cap g^*(X'_\Gamma)$.

Suppose that $w_{DC} \in g^*(X_\Gamma) \cap g^*(X'_\Gamma)$ for an arbitrary $w_{DC} \in W_{DC}$. By the definition of g^*, there must be some $x_\Gamma \in X_\Gamma$ such that $f(w_{DC}) = x_\Gamma$ and $x'_\Gamma \in X'_\Gamma$ such that $f(w_{DC}) = x'_\Gamma$. Since f is a function, $x_\Gamma = x'_\Gamma$ and $x_\Gamma \in X_\Gamma \cap X'_\Gamma$. By the definition of g^* again, $w_{DC} \in g^*(X_\Gamma \cap X'_\Gamma)$. So, $g^*(X_\Gamma) \cap g^*(X'_\Gamma) \subseteq g^*(X_\Gamma \cap X'_\Gamma)$.

6: Suppose that $w_{DC} \in g^*(X_\Gamma)$ for an arbitrary $w_{DC} \in W_{DC}$. By the definition of g^*, there must be some $x_\Gamma \in X_\Gamma$ such that $f(w_{DC}) = x_\Gamma$. By the definition of f, $x_\Gamma \subseteq w_{DC}$. Since w_{DC} is a maximally consistent set of DC, $\varphi_{x_\Gamma} \in w_{DC}$. Hence, $\varphi_{X_\Gamma} \in w_{DC}$. So, $g^*(X_\Gamma) \subseteq \left|\varphi_{X_\Gamma}\right|_{DC}$.

Suppose that $w_{DC} \in \left|\varphi_{X_\Gamma}\right|_{DC}$ for an arbitrary $w_{DC} \in W_{DC}$. Then $\varphi_{X_\Gamma} \in w_{DC}$. Since w_{DC} is a maximally consistent set of DC, there must be some $x_\Gamma \in X_\Gamma$ such that $\varphi_{x_\Gamma} \in w_{DC}$. Hence, $x_\Gamma \subseteq w_{DC}$ and $f(w_{DC}) = x_\Gamma$. Thus $w_{DC} \in g^*(X_\Gamma)$. So, $\left|\varphi_{X_\Gamma}\right|_{DC} \subseteq g^*(X_\Gamma)$.

7: It follows from item 3 and 6.

8: By the definition of g^*, it is trivially true that $g^*(W_\Gamma) \subseteq W_{DC}$. On the other hand, by definition of f, $g(W_{DC}) \subseteq W_\Gamma$. By item 2, $g^*(g(W_{DC})) \subseteq g^*(W_\Gamma)$. By item 4, $W_{DC} \subseteq g^*(g(W_{DC}))$. Thus, $W_{DC} \subseteq g^*(W_\Gamma)$.

$g^*(W_\Gamma) = |\varphi_{W_\Gamma}|_{DC}$ follows from item 6.

9: It follows from item 5 and 6.

10: Let $X_\Gamma = |\alpha|_\Gamma$. By Proposition 6(2), $\left|\varphi_{|\alpha|_\Gamma}\right|_{DC} = |\alpha|_{DC}$. Thus, $\left|\varphi_{X_\Gamma}\right|_{DC} = |\alpha|_{DC}$. By item 6, $g^*(X_\Gamma) = \left|\varphi_{X_\Gamma}\right|_{DC}$. So, $g^*(|\alpha|_\Gamma) = |\alpha|_{DC}$. $g^*(|\alpha|_\Gamma) = \left|\varphi_{|\alpha|_\Gamma}\right|_{DC} = |\alpha|_{DC}$.

11: By item 10, $g(|\alpha|_{DC}) = g(g^*(|\alpha|_\Gamma))$. By item 3, $g(g^*(|\alpha|_\Gamma)) = |\alpha|_\Gamma$. So, $g(|\alpha|_{DC}) = |\alpha|_\Gamma$.

12: By item 11, $g^*(g(|\alpha|_{DC})) = g^*(|\alpha|_\Gamma)$. By item 10, $g^*(|\alpha|_\Gamma) = |\alpha|_{DC}$. So, $g^*(g(|\alpha|_{DC})) = |\alpha|_{DC}$. □

Definition 19. *Given a set Γ of formulas, we can construct a frame $\mathfrak{F}_\Gamma = \langle W_\Gamma, \circledast_\Gamma \rangle$, where $\circledast_\Gamma$ is a function $\wp(W_\Gamma) \times \wp(W_\Gamma) \longrightarrow \wp(W_\Gamma)$, satisfying: for any $X_\Gamma, Y_\Gamma \subseteq W_\Gamma$, $\circledast_\Gamma(X_\Gamma, Y_\Gamma) = g(\circledast_{DC}(|\varphi_{X_\Gamma}|_{DC}, |\varphi_{Y_\Gamma}|_{DC}))$.*

Lemma 2. $[X_\Gamma, Y_\Gamma]_\Gamma = g([|\varphi_{X_\Gamma}|_{DC}, |\varphi_{Y_\Gamma}|_{DC}]_{DC})$

Proof. Show $[X_\Gamma, Y_\Gamma]_\Gamma \subseteq g([|\varphi_{X_\Gamma}|_{DC}, |\varphi_{Y_\Gamma}|_{DC}]_{DC})$. For an arbitrary $w_\Gamma \in [X_\Gamma, Y_\Gamma]_\Gamma$, there exists Z_Γ such that $w_\Gamma \in Z_\Gamma$ and $\circledast_\Gamma(Z_\Gamma, X_\Gamma) \subseteq Y_\Gamma$. Thus, $\circledast_\Gamma(\{w_\Gamma\}, X_\Gamma) \subseteq \circledast_\Gamma(Z_\Gamma, X_\Gamma) \subseteq Y_\Gamma$. $g(\circledast_{DC}(\left|\varphi_{\{w_\Gamma\}}\right|_{DC}, |\varphi_{X_\Gamma}|_{DC})) \subseteq Y_\Gamma$, by definition of $\circledast_\Gamma$. By Proposition 7(2,4,10), $\circledast_{DC}(\left|\varphi_{\{w_\Gamma\}}\right|_{DC}, |\varphi_{X_\Gamma}|_{DC}) \subseteq g^*(g(\circledast_{DC}(\left|\varphi_{\{w_\Gamma\}}\right|_{DC}, |\varphi_{X_\Gamma}|_{DC}))) \subseteq g^*(Y_\Gamma) = |\varphi_{Y_\Gamma}|_{DC}$. By Theorem 5, $\left|\varphi_{\{w_\Gamma\}}\right|_{DC} \subseteq |\varphi_{X_\Gamma} > \varphi_{Y_\Gamma}|_{DC}$. By Theorem 6, $\left|\varphi_{\{w_\Gamma\}}\right|_{DC} \subseteq [|\varphi_{X_\Gamma}|_{DC}, |\varphi_{Y_\Gamma}|_{DC}]_{DC}$. By Proposition 7(1), $g(\left|\varphi_{\{w_\Gamma\}}\right|_{DC}) \subseteq g([|\varphi_{X_\Gamma}|_{DC}, |\varphi_{Y_\Gamma}|_{DC}]_{DC})$. By Proposition 7(11), $w_\Gamma \in g(\left|\varphi_{\{w_\Gamma\}}\right|_{DC}) \subseteq g([|\varphi_{X_\Gamma}|_{DC}, |\varphi_{Y_\Gamma}|_{DC}]_{DC})$.

Show $g([|\varphi_{X_\Gamma}|_{DC}, |\varphi_{Y_\Gamma}|_{DC}]_{DC}) \subseteq [X_\Gamma, Y_\Gamma]_\Gamma$. For any w_Γ, if w_Γ is an element of $g([|\varphi_{X_\Gamma}|_{DC}, |\varphi_{Y_\Gamma}|_{DC}]_{DC})$, then there exists $Z_{DC} \subseteq W_{DC}$ such that $\circledast_{DC}(Z_{DC}, |\varphi_{X_\Gamma}|_{DC}) \subseteq |\varphi_{Y_\Gamma}|_{DC}$ and $w_\Gamma \in g(Z_{DC})$. By Proposition 7(1, 11), $g(\circledast_{DC}(Z_{DC}, |\varphi_{X_\Gamma}|_{DC})) \subseteq g(|\varphi_{Y_\Gamma}|_{DC}) = Y_\Gamma$. By definition of $\circledast_\Gamma$, $\circledast_\Gamma(g(Z_{DC}), X_\Gamma) = g(\circledast_{DC}(Z_{DC}, |\varphi_{X_\Gamma}|_{DC})) \subseteq Y_\Gamma$. Since $w_\Gamma \in g(Z_{DC})$, $\circledast_\Gamma(\{w_\Gamma\}, X_\Gamma) \subseteq \circledast_\Gamma(g(Z_{DC}), X_\Gamma) \subseteq Y_\Gamma$. By definition of $[X_\Gamma, Y_\Gamma]_\Gamma$, $\{w_\Gamma\} \subseteq [X_\Gamma, Y_\Gamma]_\Gamma$. Hence, $w_\Gamma \in [X_\Gamma, Y_\Gamma]_\Gamma$. □

Lemma 3. $\left|\varphi_{[X_\Gamma, Y_\Gamma]_\Gamma}\right|_{DC} = |\varphi_{X_\Gamma} > \varphi_{Y_\Gamma}|_{DC}$

Proof. By Lemma 2, $[X_\Gamma, Y_\Gamma]_\Gamma = g([|\varphi_{X_\Gamma}|_{DC}, |\varphi_{Y_\Gamma}|_{DC}]_{DC})$. By Theorem 6, $[|\varphi_{X_\Gamma}|_{DC}, |\varphi_{Y_\Gamma}|_{DC}]_{DC} = |\varphi_{X_\Gamma} > \varphi_{Y_\Gamma}|_{DC}$. $g^*([X_\Gamma, Y_\Gamma]_\Gamma) = g^*(g([|\varphi_{X_\Gamma}|_{DC}, |\varphi_{Y_\Gamma}|_{DC}]_{DC})) = g^*(g(|\varphi_{X_\Gamma} > \varphi_{Y_\Gamma}|_{DC}))$. $g^*([X_\Gamma, Y_\Gamma]_\Gamma) =$

$\left|\varphi_{[X_\Gamma,Y_\Gamma]\Gamma}\right|_{DC}$ and $g^*(g(|\varphi_{X_\Gamma} > \varphi_{Y_\Gamma}|_{DC})) = |\varphi_{X_\Gamma} > \varphi_{Y_\Gamma}|_{DC}$, by Proposition 7(10,12). Therefore, $\left|\varphi_{[X_\Gamma,Y_\Gamma]\Gamma}\right|_{DC} = |\varphi_{X_\Gamma} > \varphi_{Y_\Gamma}|_{DC}$. □

Theorem 8. *$\circledast_\Gamma$ satisfies conditions 2(a)-2(e) in Definition 3.*

Proof. 2(a): Let us show that, for any X_Γ, X'_Γ, $Y_\Gamma \subseteq W_\Gamma$, if $X_\Gamma \subseteq X'_\Gamma$ then $\circledast_\Gamma(X_\Gamma, Y_\Gamma) \subseteq \circledast_\Gamma(X'_\Gamma, Y_\Gamma)$. Suppose that $X_\Gamma \subseteq X'_\Gamma$. Then $|\varphi_{X_\Gamma}|_{DC} \subseteq \left|\varphi_{X'_\Gamma}\right|_{DC}$. By Theorem 7(1), $\circledast_{DC}(|\varphi_{X_\Gamma}|_{DC}, |\varphi_{Y_\Gamma}|_{DC}) \subseteq \circledast_{DC}(\left|\varphi_{X'_\Gamma}\right|_{DC}, |\varphi_{Y_\Gamma}|_{DC})$. $g(\circledast_{DC}(|\varphi_{X_\Gamma}|_{DC},|\varphi_{Y_\Gamma}|_{DC})) \subseteq g(\circledast_{DC}(\left|\varphi_{X'_\Gamma}\right|_{DC}, |\varphi_{Y_\Gamma}|_{DC}))$, by Proposition 7(1). Then, $\circledast_\Gamma(X_\Gamma, Y_\Gamma) \subseteq \circledast_\Gamma(X'_\Gamma, Y_\Gamma)$, by the definition of $\circledast_\Gamma$.

2(b): We show that, for any X_Γ, Y_Γ, $Z_\Gamma \subseteq W_\Gamma$, if $\circledast_\Gamma(\{w_\Gamma\}, Y_\Gamma) \subseteq Z_\Gamma$ for every $w_\Gamma \in X_\Gamma$, then $\circledast_\Gamma(X_\Gamma, Y_\Gamma) \subseteq Z_\Gamma$. Suppose that $\circledast_\Gamma(\{w_\Gamma\},Y_\Gamma) \subseteq Z_\Gamma$ for every $w_\Gamma \in X_\Gamma$. By the definition of $\circledast_\Gamma$ and Proposition 7(2), $g^*(\circledast_\Gamma(\{w_\Gamma\},Y_\Gamma)) \subseteq g^*(Z_\Gamma)$ for every $w_\Gamma \in X_\Gamma$. By the definition of $\circledast_\Gamma$ and Proposition 7(4,6), for every $w_\Gamma \in X_\Gamma$, $\circledast_{DC}(\left|\varphi_{\{w_\Gamma\}}\right|_{DC}, |\varphi_{Y_\Gamma}|_{DC}) \subseteq g^*(g(\circledast_{DC}(\left|\varphi_{\{w_\Gamma\}}\right|_{DC}, |\varphi_{Y_\Gamma}|_{DC}))) = g^*(\circledast_\Gamma(\{w_\Gamma\},Y_\Gamma)) \subseteq g^*(Z_\Gamma) = |\varphi_{Z_\Gamma}|_{DC}$. For any $w_{DC} \in |\varphi_{X_\Gamma}|_{DC}$, there must be some $w'_\Gamma \in X_\Gamma$ such that $w_{DC} \in \left|\varphi_{\{w'_\Gamma\}}\right|_{DC}$. By Theorem 7(1), $\circledast_{DC}(\{w_{DC}\}, |\varphi_{Y_\Gamma}|_{DC}) \subseteq \circledast_{DC}(\left|\varphi_{\{w_{\Gamma'}\}}\right|_{DC}, |\varphi_{Y_\Gamma}|_{DC}) \subseteq |\varphi_{Z_\Gamma}|_{DC}$. By Theorem 7(2), we have $\circledast_{DC}(|\varphi_{X_\Gamma}|_{DC}, |\varphi_{Y_\Gamma}|_{DC}) \subseteq |\varphi_{Z_\Gamma}|_{DC}$. $g(\circledast_{DC}(|\varphi_{X_\Gamma}|_{DC},|\varphi_{Y_\Gamma}|_{DC})) \subseteq g(|\varphi_{Z_\Gamma}|_{DC}] = g(g^*(Z_\Gamma))$, by Proposition 7(1). $\circledast_\Gamma(X_\Gamma, Y_\Gamma) \subseteq Z_\Gamma$, by the definition of $\circledast_\Gamma$ and Proposition 7(3).

2(c): It can be shown that, for any X_Γ, Y_Γ, $Z_\Gamma \subseteq W_\Gamma$, if $\circledast_\Gamma(X_\Gamma, Y_\Gamma) \subseteq Z_\Gamma$ then $\circledast_\Gamma(W_\Gamma, X_\Gamma \cap Y_\Gamma) \subseteq Z_\Gamma$. Suppose that $\circledast_\Gamma(X_\Gamma, Y_\Gamma) \subseteq Z_\Gamma$. $\circledast_{DC}(|\varphi_{X_\Gamma}|_{DC}, |\varphi_{Y_\Gamma}|_{DC}) \subseteq |\varphi_{Z_\Gamma}|_{DC}$, by the definition of $\circledast_\Gamma$ and Proposition 7(2,4,6). By Theorem 7(3), we have $\circledast_{DC}(W_{DC}, |\varphi_{X_\Gamma}|_{DC} \cap |\varphi_{Y_\Gamma}|_{DC}) \subseteq |\varphi_{Z_\Gamma}|_{DC}$. Since $W_{DC} = |\varphi_{W_\Gamma}|_{DC}$ and $|\varphi_{X_\Gamma}|_{DC} \cap |\varphi_{Y_\Gamma}|_{DC} = |\varphi_{X_\Gamma \cap Y_\Gamma}|_{DC}$, we have $\circledast_{DC}(|\varphi_{W_\Gamma}|_{DC}, |\varphi_{X_\Gamma \cap Y_\Gamma}|_{DC}) \subseteq g^*(Z_\Gamma)$. By the definition of $\circledast_\Gamma$ and Proposition 7(3,1), $\circledast_\Gamma(W_\Gamma, X_\Gamma \cap Y_\Gamma) \subseteq Z_\Gamma$.

2(d): Given any X_Γ, $Y_\Gamma \subseteq W_\Gamma$, $\circledast_{DC}(|\varphi_{X_\Gamma}|_{DC},|\varphi_{Y_\Gamma}|_{DC}) \subseteq |\varphi_{Y_\Gamma}|_{DC}$, by Theorem 7(4). By Proposition 7(1,7), $g(\circledast_{DC}(|\varphi_{X_\Gamma}|_{DC},|\varphi_{Y_\Gamma}|_{DC})) \subseteq g(|\varphi_{Y_\Gamma}|_{DC}) = g(g^*(Y_\Gamma)) = Y_\Gamma$. By the definition of $\circledast_\Gamma$, $\circledast_\Gamma(X_\Gamma,Y_\Gamma) \subseteq Y_\Gamma$.

2(e): Given any X_Γ, Y_Γ, Y'_Γ, $Z_\Gamma \subseteq W_\Gamma$, suppose that $\circledast_\Gamma(X_\Gamma,Y_\Gamma \cap Y'_\Gamma) \subseteq Z_\Gamma$. According to the definition of $\circledast_\Gamma$, we have $g(\circledast_{DC}(|\varphi_{X_\Gamma}|_{DC}, \left|\varphi_{Y_\Gamma \cap Y'_\Gamma}\right|_{DC})) \subseteq Z_\Gamma$. By Proposition 7(2,4,6,9), $\circledast_{DC}(|\varphi_{X_\Gamma}|_{DC}, \left|\varphi_{Y_\Gamma \cap Y'_\Gamma}\right|_{DC}) = \circledast_{DC}(|\varphi_{X_\Gamma}|_{DC}, |\varphi_{Y_\Gamma}|_{DC} \cap \left|\varphi_{Y'_\Gamma}\right|_{DC}) \subseteq |\varphi_{Z_\Gamma}|_{DC}$. Then, $\circledast_{DC}(|\varphi_{X_\Gamma}|_{DC}, |\varphi_{Y_\Gamma}|_{DC} \cap \left|\varphi_{Y_\Gamma} > \varphi_{Y'_\Gamma}\right|_{DC} \cap \left|\varphi_{Y'_\Gamma} > \varphi_{Z_\Gamma}\right|_{DC}) \subseteq |\varphi_{Z_\Gamma}|_{DC}$, by Theorem 7(5). $\circledast_{DC}(|\varphi_{X_\Gamma}|_{DC},$

$\left|\varphi_{Y_\Gamma \cap [Y_\Gamma, Y'_\Gamma] \cap [Y'_\Gamma, Z_\Gamma]}\right|_{DC}) \subseteq |\varphi_{Z_\Gamma}|_{DC}$, by Lemma 3. $g(\circledast_{DC}(|\varphi_{X_\Gamma}|_{DC}, |\varphi_{Y_\Gamma}|_{DC} \cap \left|\varphi_{[Y_\Gamma, Y'_\Gamma]}\right|_{DC} \cap \left|\varphi_{[Y'_\Gamma, Z_\Gamma]}\right|_{DC})) \subseteq g(|\varphi_{Z_\Gamma}|_{DC})$, by Proposition 7(1,9). By Definition the definition of $\circledast_\Gamma$ and Proposition 7(7), $\circledast_\Gamma(X_\Gamma, Y_\Gamma \cap [Y_\Gamma, Y'_\Gamma] \cap [Y'_\Gamma, Z_\Gamma]) \subseteq Z_\Gamma$. □

Lemma 4. *Given any finite set Γ of formulas, formulas α and β, and $X_\Gamma \subseteq W_\Gamma$, $\circledast_\Gamma(X_\Gamma, |\alpha|_\Gamma) \subseteq |\beta|_\Gamma$ iff $\circledast_{DC}(|\varphi_{X_\Gamma}|_{DC}, \left|\varphi_{|\alpha|_\Gamma}\right|_{DC}) \subseteq |\beta|_{DC}$.*

Proof. Suppose that $\circledast_\Gamma(X_\Gamma, |\alpha|_\Gamma) \subseteq |\beta|_\Gamma$. $g(\circledast_{DC}(|\varphi_{X_\Gamma}|_{DC}, \left|\varphi_{|\alpha|_\Gamma}\right|_{DC})) \subseteq |\beta|_\Gamma$, by Definition 19. Then $g^*(g(\circledast_{DC}(|\varphi_{X_\Gamma}|_{DC}, \left|\varphi_{|\alpha|_\Gamma}\right|_{DC}))) \subseteq g^*(|\beta|_\Gamma)$, by Proposition 7 (2). Then $g^*(g(\circledast_{DC}(|\varphi_{X_\Gamma}|_{DC}, |\alpha|_{DC}))) \subseteq |\beta|_{DC}$, by Proposition 7 (10). Then $\circledast_{DC}(|\varphi_{X_\Gamma}|_{DC}, |\alpha|_{DC}) \subseteq |\beta|_{DC}$, by Proposition 7 (4).

To prove the other direction of inclusion, suppose that $\circledast_{DC}(|\varphi_{X_\Gamma}|_{DC}, |\alpha|_{DC}) \subseteq |\beta|_{DC}$. By Proposition 7 (1), $g(\circledast_{DC}(|\varphi_{X_\Gamma}|_{DC}, \left|\varphi_{|\alpha|_\Gamma}\right|_{DC})) \subseteq g(|\beta|_{DC})$. Then $\circledast_\Gamma(X_\Gamma, |\alpha|_\Gamma) \subseteq |\beta|_\Gamma$, by Definition 19 and Proposition 7 (10,11). □

Lemma 5. *Given any finite set Γ of formulas, formulas α and β, and $X_\Gamma \subseteq W_\Gamma$, $|\varphi_{X_\Gamma}|_{DC} \subseteq |\alpha > \beta|_{DC}$ iff $g(|\varphi_{X_\Gamma}|_{DC}) \subseteq g(|\alpha > \beta|_{DC})$.*

Proof. By Proposition 7 (2), $g(|\varphi_{X_\Gamma}|_{DC}) \subseteq g(|\alpha > \beta|_{DC})$ if and only if $g^*(g(|\varphi_{X_\Gamma}|_{DC}) \subseteq g^*(g(|\alpha > \beta|_{DC}))$. By Proposition 7 (12), $g^*(g(|\varphi_{X_\Gamma}|_{DC})) = |\varphi_{X_\Gamma}|_{DC}$ and $g^*(g(|\alpha > \beta|_{DC})) = |\alpha > \beta|_{DC}$. It follows that $g^*(g(|\varphi_{X_\Gamma}|_{DC})) \subseteq g^*(g(|\alpha > \beta|_{DC}))$ iff $|\varphi_{X_\Gamma}|_{DC} \subseteq |\alpha > \beta|_{DC}$. Therefore, $|\varphi_{X_\Gamma}|_{DC} \subseteq |\alpha > \beta|_{DC}$ iff $g(|\varphi_{X_\Gamma}|_{DC}) \subseteq g(|\alpha > \beta|_{DC})$. □

Theorem 9. *Given a set Γ of formulas, we can construct a model $\mathfrak{M}_\Gamma = \langle W_\Gamma, \circledast_\Gamma \sigma_\Gamma \rangle$, where σ_Γ is an assignment function satisfying: for each propositional variable p, $\sigma_\Gamma(p) = \{w_\Gamma \in W_\Gamma \mid p \in w_\Gamma\}$. Then, for each $\varphi \in \mathsf{CCL}(\Gamma)$, we have $\|\varphi\|^{\mathfrak{M}_\Gamma} = |\varphi|_\Gamma$. That is, for each $X_\Gamma \subseteq W_\Gamma$, $\mathfrak{M}_\Gamma \models_{X_\Gamma} \varphi$ iff $X_\Gamma \subseteq |\varphi|_\Gamma$.*

Proof. We prove this theorem by induction on φ.

Case 1: φ is a propositional variable p. By the definition of $\|\varphi\|^{\mathfrak{M}_\Gamma}$, $\sigma_\Gamma(p)$ and $|\varphi|_\Gamma$, $\|\varphi\|^{\mathfrak{M}_\Gamma} = \|p\|_\Gamma = \sigma_\Gamma(p) = |p|_\Gamma$.

Case 2: φ is in the form of $\neg\alpha$. $\|\varphi\|^{\mathfrak{M}_\Gamma} = \|\neg\alpha\|_\Gamma = W_\Gamma - \|\alpha\|_\Gamma$. By the induction hypothesis, $\|\varphi\|^{\mathfrak{M}_\Gamma} = W_\Gamma - |\alpha|_\Gamma$. By Proposition 5(1), $\|\varphi\|^{\mathfrak{M}_\Gamma} = |\neg\alpha|_\Gamma = |\varphi|_\Gamma$.

Case 3: φ is in the form of $\alpha \to \beta$. $\|\varphi\|^{\mathfrak{M}_\Gamma} = \|\alpha \to \beta\|_\Gamma = (W_\Gamma - \|\alpha\|_\Gamma) \cup \|\beta\|_\Gamma$. By the induction hypothesis, $\|\varphi\|^{\mathfrak{M}_\Gamma} = (W_\Gamma - |\alpha|_\Gamma) \cup |\beta|_\Gamma$. By Proposition 5(2), $\|\varphi\|^{\mathfrak{M}_\Gamma} = |\alpha \to \beta|_\Gamma = |\varphi|_\Gamma$.

Case 4: φ is in the form of $\alpha > \beta$. $\|\varphi\|^{\mathfrak{M}_\Gamma} = \|\alpha > \beta\|_\Gamma = \bigcup\{X_\Gamma \subseteq W_\Gamma \mid \circledast_\Gamma(X_\Gamma, \|\alpha\|_\Gamma) \subseteq \|\beta\|_\Gamma\}$. By the induction hypothesis, $\|\varphi\|^{\mathfrak{M}_\Gamma} = \bigcup\{X_\Gamma \subseteq W_\Gamma \mid \circledast_\Gamma(X_\Gamma, |\alpha|_\Gamma) \subseteq |\beta|_\Gamma\}$. By Lemma 4, $\|\varphi\|^{\mathfrak{M}_\Gamma} = \bigcup\{X_\Gamma \subseteq W_\Gamma \mid \circledast_{DC}(|\varphi_{X_\Gamma}|_{DC}, |\alpha|_{DC}) \subseteq |\beta|_{DC}\}$. By Theorem 5, $\|\varphi\|^{\mathfrak{M}_\Gamma} = \bigcup\{X_\Gamma \subseteq W_\Gamma \mid |\varphi_{X_\Gamma}|_{DC} \subseteq |\alpha > \beta|_{DC}\}$. By Lemma 5, $\|\varphi\|^{\mathfrak{M}_\Gamma} = \bigcup\{X_\Gamma \subseteq W_\Gamma \mid g(|\varphi_{X_\Gamma}|_{DC}) \subseteq g(|\alpha > \beta|_{DC})\} = \bigcup\{X_\Gamma \subseteq W_\Gamma \mid X_\Gamma \subseteq |\alpha > \beta|_\Gamma\} = |\alpha > \beta|_\Gamma = |\varphi|_\Gamma$. □

The completeness theorem (Theorem 4) follows from the above Theorem 8 and Theorem 9.

What we have proved is the frame-completeness. That is, the system DC is complete with respect to the class of all $\mathfrak{L}_>$-frames. This result is stronger than the model-completeness that was often proved for some systems of defaults like in Delgrande's work ([8]). Since the small model we construct to falsify a given unprovable formula is finite, the finite model property and decidability of the system DC naturally follow.

5 What to give up

Given a premise set Γ, the system DC enables us to deduce all default conclusions of Γ. These default conclusions are intermediate conclusions of Γ, and are not yet warranted. When a set Γ of premises is not itself inconsistent, its default conclusions may be contradictory. To avoid contradictions to be warranted as conclusions of Γ, we will have to give up some default conclusions while retaining more preferable ones. In this section we discuss which default conclusions to give up in face of contradiction, in order to obtain warranted conclusions.

5.1 Definition of $\mid\!\sim$

Definition 20. *Given a set Γ of formulas, $\succeq_\Gamma$ denotes a transitive binary relation on Γ. Let $\alpha \succ_\Gamma \beta$ ($\alpha, \beta \in \Gamma$) be an abbreviation for ($\alpha \succeq_\Gamma \beta$) and $\neg(\beta \succeq_\Gamma \alpha)$.*

It is easy to see that $\succ_\Gamma$ is irreflexive, asymmetric and transitive. $\succ_\Gamma$ can be understood as a relation of "more preferable than". When the domain Γ of the relation $\succeq_\Gamma$ has been clearly stated in a context and no confusion will arise, the subscript may be omitted.

Definition 21. *Given a set Γ of formulas and a relation $\succeq$ defined on it, let $\trianglerighteq$ be a binary relation on $\wp(\Gamma)$, satisfying: $\forall\Delta, \Theta \subseteq \Gamma$, $\Delta \trianglerighteq \Theta$ iff $\exists\delta\exists\theta((\delta \in \Delta) \wedge (\theta \in \Theta) \wedge (\delta \succ \theta))$. If $\Delta \trianglerighteq \Theta$ and $\Theta \ntrianglerighteq \Delta$, then we say that Δ is more preferable than Θ and denote this relation by $\Delta \triangleright \Theta$.*

In daily life, people naturally tend to have preferences among statements. For example, scientific announcement is more trustworthy than folk gossip. Thus, the initial preference relation is defined among formulas as in Definition 20. However, an argument often has more than one premise. When the underlying logic remains the same, the strength of an argument is determined by the strength of its premises. To compare the strength of two arguments, it is insufficient to compare only one premise from each side. Rather, their entire premise sets have to be compared for the overall preference. Due to this consideration, the initial preference relation among formulas has to be extended to cover the preference relation among sets of formulas. Definition 21 serves this purpose. The defined relation $\rhd$ of "more preferable than" among sets of premises is asymmetric.

Proposition 8. *Let $\rhd$ be a binary relation defined on $\wp(\Gamma)$, as stated in Definition 21. For all $\forall\Delta,\Theta \subseteq \Gamma$, if $\Delta \rhd \Theta$ then $\Theta \not\rhd \Delta$.*

For a default conclusion α, it could be the case that it is simultaneously supported by several local arguments. Similarly, there may also be several arguments locally supporting its rival conclusion $\neg\alpha$. In order to determine which default conclusion should eventually come out as the global conclusion, two groups of arguments have to be compared for their strength. This can be done through comparing two groups of premise sets, with everything else being equal. Thus, there is a need to further define a preference relation $\geqslant$ among sets of premise sets in terms of $\rhd$. The domain of the relation $\geqslant$ is now $\wp(\wp(\Gamma))$, which is on a higher level than $\wp(\Gamma)$.

Definition 22. *Given a binary relation $\rhd$ defined on $\wp(\Gamma)$ via $\succeq_\Gamma$, let $\geqslant$ be a binary relation on $\wp(\wp(\Gamma))$, satisfying: $\forall\Phi,\Psi \subseteq \wp(\Gamma)$, $\Phi \geqslant \Psi$ iff $\forall\Theta((\Theta \in \Psi) \rightarrow \exists\Delta((\Delta \in \Phi) \wedge (\Delta \rhd \Theta)))$. If $\Phi \geqslant \Psi$ and $\Psi \not\geqslant \Phi$, then we say that Φ is more preferable than Ψ and denote this relation by $\Phi > \Psi$.*

The relation $>$ so defined is asymmetric, as stated in the Proposition 9.

Proposition 9. *Let $>$ be a binary relation defined on $\wp(\wp(\Gamma))$ as stated in Definition 22. For all $\forall\Phi,\Psi \subseteq \wp(\Gamma)$, if $\Phi > \Psi$ then $\Psi \not\geqslant \Phi$.*

From a mathematical point of view, there are many ways of obtaining an ordering over $\wp(\Gamma)$ and $\wp(\wp(\Gamma))$ on the basis of an ordering over Γ. Thus, it is not immediately obvious which among them are the most appropriate for these purposes. Our choice of the definitions is guided and corrected in our trials by seeing their results.

Definition 23. *Let Γ be a set of formulas. We say that Γ is consistent iff there is no formula α such that $\alpha \in \Gamma$ and $\neg\alpha \in \Gamma$.*

Definition 24. *Let* $\Delta = \{\alpha_1, \alpha_2, ..., \alpha_n\}$ *be a non-empty finite set of formulas.* $\wedge\Delta = \alpha_1 \wedge \alpha_2 \wedge ... \wedge \alpha_n$, $n \geqslant 1$

Definition 25. *Given a finite set* Δ *of formulas and a formal logic system* $S(L)$, $C_n(\Delta) = \{\alpha \mid\vdash_{S(L)} (\wedge\Delta) > \alpha\}$, *and*

$$C_N(\Delta) = \begin{cases} C_n(\Delta), \text{ if } C_n(\Delta) \text{ is consistent} \\ \varnothing, \text{ otherwise} \end{cases}.$$

Definition 26. *Given a set* Γ *of formulas and a formal logic system* $S(L)$, $C_M(\Gamma) = \{\alpha \mid \exists\Delta \subseteq \Gamma, \quad \vdash_{S(L)} (\wedge\Delta) \to \alpha\}$.

The elements of $C_M(\Gamma)$ are called *deductive consequences* of Γ. If the Deduction Theorem holds in $S(L)$ (i.e., $\vdash_{S(L)} (\wedge\Delta) \to \alpha$ iff $\wedge\Delta \vdash_{S(L)} \alpha$), then $C_M(\Gamma)$ is the deductive closure of Γ in logic $S(L)$. The operation of deductive closure is idempotent, that is, $C_M(\Gamma) = C_M(C_M(\Gamma))$. Moreover, for any $\Gamma \subseteq \Gamma' \subseteq C_M(\Gamma)$, $C_M(\Gamma) = C_M(\Gamma')$.

Definition 27. *Given a set* Γ *of formulas, let* $\mathsf{CCL}_M(\Gamma) = \mathsf{CCL}(\Gamma) \cap C_M(\Gamma)$. $\mathsf{CCL}_M(\Gamma)$ *is the set of all components of* Γ *that are also deductive consequences of* Γ.

It is easy to see that $\Gamma \subseteq \mathsf{CCL}_M(\Gamma) \subseteq C_M(\Gamma)$. If Γ is finite, so is $\mathsf{CCL}_M(\Gamma)$. The intention of defining $\mathsf{CCL}_M(\Gamma)$ is to have a finite version of some "relevant" deductive consequences of Γ. For instance, all theorems of $S(L)$ will be included in $C_M(\Gamma)$, but they are not relevant to Γ in the sense that they are derivable in $S(L)$ with or without Γ. On the other hand, if $\alpha, \alpha \to \beta \in \Gamma$, we would like to have $\beta \in \mathsf{CCL}_M(\Gamma)$. Putting it in metaphorical words, $\mathsf{CCL}_M(\Gamma)$ disassembles Γ into its components, catches those that are previously embedded as consequents of material implications in Γ, and then brings them to the surface for future use. The set $\mathsf{CCL}_M(\Gamma)$ retains "useful" deductive consequences of Γ like β to our task. In the meantime, the size of $\mathsf{CCL}_M(\Gamma)$ is cut down to finite. This makes $\mathsf{CCL}_M(\Gamma)$ more manageable than $C_M(\Gamma)$ for future use.

Definition 28. *Given a set* Δ *of formulas and a formula* α, Δ *is called a* simplest premise set *of* α *with respect to the* $>$*-inference, if*

1. $\alpha \in C_N(\Delta)$, *and*
2. *for any* $\Theta \subset \mathsf{CCL}_M(\Delta)$, $\alpha \notin C_N(\Theta)$.

The first clause of the definition above ensures that Δ is a premise set of α with respect to the $>$-inference. The second clause requires Δ to be as simple as possible. The requirement of being the simplest places restrictions on Δ in two dimensions. One is that Δ does not contain more premises than needed. The type of inference based on "normally follows", unlike the material implication $\rightarrow$, can be influenced by additional premises. The notion of the simplest premise set is to rule out those premises that have no contribution to obtaining α in a $>$-inference. The other dimension is that each premise must be in a simple form like atomic formulas, their negations, and $>$-formulas. This is to prevent additional premises from hiding in complex formulas like $\rightarrow$-formulas and disjunctions. For example, let $\Delta = \{p > r, p > r \rightarrow q > \neg r, q\}$. $\neg r \in C_N(\Delta)$, but Δ is not a simplest premise set of α. Δ is equivalent to $\Delta' = \{p > r, q > \neg r, q\}$. In Δ', $p > r$ is an extra premise for the default conclusion $\neg r$. Let $\Theta = \{q > \neg r, q\} \subset \mathsf{CCL}_M(\Delta)$. Θ is a simplest premise set of $\neg r$.

For the $\rightarrow$ related inference, there is no need to talk about the simplest premise set, since $\rightarrow$-inference is monotonic and does not "fear" additional premises. For simplicity, we will hereafter omit the phrase of "with respect to the $>$-inference", and just say "Δ is a simplest premise set of α" or "Δ is not a simplest premise set of α".

Definition 29. *For any formula α, a set Γ of formulas, and a formal logic system $S(L)$, let $SPS_\Gamma(\alpha) = \{\Delta \mid \Delta$ is a simplest premise set of α, and $\Delta \subseteq \mathsf{CCL}_M(\Gamma)\}$ denote all Γ-relevant simplest premise sets of α.*

$SPS_\Gamma(\alpha)$ is the set of all simplest premise sets of α that consist of deductive consequences of Γ.

Definition 30. *Given a finite set Γ of formulas and a transitive binary relation $\succeq$ on $\mathsf{CCL}_M(\Gamma)$, and a formula α,*

$\Gamma \mid\!\sim_{S(L),\succeq} \alpha$ iff

1. *$\alpha \in C_M(\Gamma)$, or*
2. *the following three conditions are satisfied:*
   - **(2a)** *$\neg\alpha \notin C_M(\Gamma)$;*
   - **(2b)** *$SPS_\Gamma(\alpha) \neq \varnothing$;*
   - **(2c)** *if $SPS_\Gamma(\neg\alpha) \neq \varnothing$, then $SPS_\Gamma(\alpha) \gtrdot SPS_\Gamma(\neg\alpha)$.*

Let $CN(\Gamma) = \{\alpha \mid \Gamma \mid\!\sim_{S(L),\succeq} \alpha\}$. $CN(\Gamma)$ denotes the set of all global conclusions of Γ with respect to $S(L)$ and $\succeq$.

The first disjunct in the above definition includes all deductive consequences of Γ as its global conclusions. Logicians are inclined to give higher preference to deductive conclusions than to default conclusions. By and large, they require if $\Gamma \vdash \alpha$ then $\Gamma \mid\!\sim \alpha$, but not vice versa. As $\vdash$ and $\mid\!\sim$ are defined with respect to a certain logic system, there could be variations of this requirement. For example, anything that can be deductively inferred by Γ in PC (or in First Order Logic, or in the same logic upon which $\mid\!\sim$ is defined) should also be nonmonotonically inferred. Definition 30 guarantees that $C_M(\Gamma) \subseteq CN(\Gamma)$.

The second disjunct expands the boundary of $CN(\Gamma)$ to include some of default conclusions as well. There are three conditions a default conclusion α must satisfy in order to be included in $CN(\Gamma)$ as a global conclusion. First, α does not conflict with a deductive consequence of Γ. Second, there exists some simplest premise set of α that infers it as a default conclusion. At last, if there are premise sets that infer $\neg\alpha$, then the group of premise sets of α must have higher rank than the group of premise sets of $\neg\alpha$. The last condition lays out the criterion for resolving conflicting default conclusions. A default conclusion α may well be a local conclusion of several simplest premise sets respectively. Similarly, its rival $\neg\alpha$ may also be backed up by a number of simplest premise sets from different aspects. To determine which one of the conflicting default conclusions should be kept as a global conclusion, comparing one premise set from each side will not be sufficient. Instead, the entire groups of the simplest premise sets from both sides should be compared for preference. Namely, the default conclusion that is overall supported by the more preferable reasons should win the battle. That is why the initial preference relation among premises has been lifted two levels high up via Definition 21 and Definition 22. The derived preference relation among groups of premise sets is needed for conflict resolution in this definition.

The first condition reflects the deductive-consequence-first principle. Recall that the characterizing feature of reasoning with defaults is that the conclusions are defeasible. When should a default conclusion be retracted? The most straightforward answer is "in the light of new evidence". It is often said that if new evidence conflicts with the previous conclusion, then the new evidence wins out and the previous conclusion should be discarded. This describes people's preference for firm facts over "soft" default conclusions. We tend to yield the latter to the former. On the one hand, given facts are deductive consequences of their own. On the other hand, deductive consequences of given facts are as reliable as facts themselves. The deductive-consequence-first principle, covering fact-first principle, is a more broad principle.

In the second and third conditions, the notion of simplest premise set is used. The requirement of simplest premise sets is to eliminate side effects from additional premises. These premises may block α to be inferred as a global conclusion by altering the preference relation between two premise sets Δ and Θ. For example, suppose that $\Gamma = \{p, p > q, r, r > \neg q, s\}$, and that $\succeq = \{\langle p, r\rangle, \langle s, p\rangle\}$. Let $\Delta = \{p, p > q\}$ and $\Theta = \{r, r > \neg q\}$, and they are simplest premise sets. Then, $q \in C_N(\Delta)$, $\neg q \in C_N(\Theta)$, and $\Delta \rhd \Theta$. So, $q \in CN(\Gamma)$. If it were not required that Θ must be a simplest premise set, then Θ could be $\{r, r > \neg q, s\}$. As a result, $\Delta \not\rhd \Theta$, and now $q \notin CN(\Gamma)$. This blocking effect caused by irrelevant premises is unwelcome and must be excluded. The requirement of simplest premise sets matches with the two-phase framework. The phase of inferring default conclusions only focuses on needed premises. This phase should not be distracted by additional premises. The defeasibility of default conclusions caused by additional premises is to be managed in the second phase via preference relations.

Definition 30 uses the relation $\succeq$ that is defined on $\mathsf{CCL}_M(\Gamma)$. Note that it is defined neither on Γ nor on $C_M(\Gamma)$. There are two reasons for constructing an intermediate set $\mathsf{CCL}_M(\Gamma)$ and taking its preference relation in Definition 30. One reason is to manage examples like this: Let $\Gamma = \{p, q, s, s \to p > q, p > \neg r, q > r\}$. Then ask whether $\neg r \in CN(\Gamma)$. Γ is very similar to the premise set of the Tweety bird example (example (9) in Section 1), except that $p > q \in \mathsf{CCL}_M(\Gamma)$ but $p > q \notin \Gamma$. That is why we must make use of the preference relation on $\mathsf{CCL}_M(\Gamma)$. Another reason is that $\mathsf{CCL}_M(\Gamma)$ is finite, while $C_M(\Gamma)$ is not. The check-up of condition (2c) in Definition 30 can be well under control if the preference relation is defined on the finite set $\mathsf{CCL}_M(\Gamma)$.

Theorem 10. *If $C_M(\Gamma)$ is consistent, then $CN(\Gamma)$ is consistent.*

Proof. Suppose that there is a formula α_0 such that both $\alpha_0 \in CN(\Gamma)$ and $\neg\alpha_0 \in CN(\Gamma)$. In the case that $\alpha_0 \in CN(\Gamma)$ because $\alpha_0 \in C_M(\Gamma)$: Since $C_M(\Gamma)$ is consistent, $\neg\alpha_0 \notin C_M(\Gamma)$; Also, $\neg\alpha_0$ cannot be a member of $CN(\Gamma)$ by going through the second disjunct of Definition 30 because condition (2a) is not satisfied. Thus, $\neg\alpha_0 \notin CN(\Gamma)$. Similarly, in the case that $\neg\alpha_0 \in CN(\Gamma)$ because $\neg\alpha_0 \in C_M(\Gamma)$, $\alpha_0 \notin CN(\Gamma)$. Now let us consider the case where both $\alpha_0 \notin C_M(\Gamma)$ and $\neg\alpha_0 \notin C_M(\Gamma)$. By Definition 30, there are simplest premise sets for α_0 and $\neg\alpha_0$, respectively. That is, $SPS_\Gamma(\alpha) \neq \varnothing$ and $SPS_\Gamma(\neg\alpha) \neq \varnothing$. In addition, Definition 30 also requires that $SPS_\Gamma(\alpha) \gtrdot SPS_\Gamma(\neg\alpha)$ and $SPS_\Gamma(\neg\alpha) \gtrdot SPS_\Gamma(\alpha)$, which contradicts Proposition 9. □

Any non-empty $C_N(\Delta)$ represents default conclusions of Γ that are locally inferred by a logic system $S(L)$ from a subset Δ of Γ. $CN(\Gamma)$ is the set of all global conclusions of Γ. It contains no more contradictory default conclusions that may come from a couple of different $C_N(\Delta)$. The definition of $\mid\sim$ is parameterized by two factors: the underlying logic system $S(L)$ and the binary relation $\succeq$. Besides deriving deductive conclusions of Γ, the system $S(L)$ is also the engine to generate candidates of default conclusions for global conclusions of Γ, and provides power to push the inference going forward. These candidates will have to go through a check-up step. Some of them will be filtered out if their opposites can be inferred from a strictly more preferable set of the simplest premise sets. The mechanism to filter out conflicting default conclusions depends on the relation $\succeq$ over $\mathsf{CCL}_M(\Gamma)$. Different constraints imposed on the binary relation $\succeq$ will affect the ways that conflicts get resolved, and hence will result in different global conclusions. System $S(L)$ is orthogonal to the relation $\succeq$. We view that a nonmonotonic inference consists of two steps. The first step is to boldly get all default conclusions. This is handled by the system $S(L)$. The second step is to check and maintain consistency, to give up less preferable default conclusions in case of conflicts. This is done based on preferable relation $\succeq$ (and also $\succ$) over $\mathsf{CCL}_M(\Gamma)$.

According to the definitions given above, a check-up and settlement is needed *only* when there are conflicting default conclusions from some $C_N(\Delta_1)$ and $C_N(\Delta_2)$, and they compete to get into the global conclusion set $CN(\Gamma)$. The focus of our method is on the conflicting conclusions, but not on the incompatible arguments. This strategy is in contrast with Dung's argumentation systems ([10]) and inheritance semantic networks developed by Horty *et al.* ([14]).

Those theories pay attention to the construction of arguments and comparison between arguments. That an argument path is permitted is defined in terms of its initial segments being permitted together with some other conditions. It seems to us that this is an attempt to inductively construct a global argument that uses all premises from some local arguments. Such an attempt inevitably leads to a complex sequence of finding out when an argument is defeated, and when a defeated argument can be re-instated, and when a re-instated argument is defeated again. One of the basic features that we summarized at the very beginning of this paper is that the complete premise set must be taken into account when a conclusion of a default reasoning is under examination. Locally inferred conclusions do not compose globally-inferable conclusions in a manageable and predictable way. Mastering the relation between the set of locally-inferred default conclusions and the set of global conclusions amounts to the tough task of managing the nonmonotonicity itself.

Makinson and Schlechta ([18]) points out that some inheritance nets suffer problems of zombie paths and floating conclusions. Zombie paths, which are not permitted by their nets, are in a dilemma situation: they should not be completely dead and do nothing to other paths, but whatever power that they retain to counteract other paths seems too strong in some scenarios. The truth may lie in the fact that the status of a partial argument (defeated or permitted) does not really matter to the whole argument for a global conclusion. A floating conclusion is one that can be reached by two conflicting and equally strong arguments. Though researchers may have different intuition on the acceptance of floating conclusions (e.g., [15, 27]), we second the position advocated by scholars like Prakken, Makinson and Schlechta that a good theory on nonmonotonic reasoning should accept floating conclusions. Some argumentation systems that share similar notions with inheritance nets also have difficulties respecting the expected activity of zombie paths and accommodating floating conclusions.

In the two-phase treatment of default reasoning that we propose, we allow and, as a matter of fact, invite the existence of competing arguments for conflicting default conclusions. An argument will not be abandoned half way simply because it reaches a statement that conflicts with other information with a higher priority. All arguments for default conclusions are welcomed and supported in the system DC, regardless of the possibility that they may lead to conflicting default conclusions. This is very different from the inductive view of default reasoning, which attempts to get everything right in the first place in a stepwise construction of a global argument. In [20], we analyzed the root causes for zombie paths and floating conclusions and concluded that the inductive approach has pushed the analogy between default reasoning and classical deduction too far.

Inspired by [1], we guarantee the consistency of the set of final global conclusions via choosing between alternative premise sets. We think that the center of our attention should be conclusions themselves, rather than arguments that lead to them. We use the preference relation on the premise sets to measure the importance of the conclusions and to resolve the conflicts. We let the underlying system DC provide arguments for default conclusions, but do not bother to trace them in a stepwise manner. Since we encapsulate the complexity of argument paths, the zombie path problem does not apply to our method. Our strategy of drawing back incompatible conclusions always takes into account the negative impact from the default conclusions deduced by system DC, including those that are supported by zombie paths.

As to floating conclusions, we are not bothered by the tension produced when some intermediate results fall on incompatible argument paths. The question of what to give up is relevant only in the face of conflicting default conclusions. Though floating conclusions are supported by different arguments, and these arguments may use incompatible information in their intermediate steps, there are no opposite conclusions fighting against them. From this perspective, we think that floating conclusions should be accepted, and our definition of global conclusions reflects our position on this issue.

5.2 Specificity as a Preference Relation

In the previous sections, we developed system DC, which will be a particular $S(L)$ in Definition 30. The preference relation has not yet been specified in our general discussion, other than as a transitive binary relation over $\mathsf{CCL}_M(\Gamma)$. There are principles that bias our choices for final global conclusions for a given premise set in commonsense entailment. Among them, the most frequently discussed in the literature are fact-first principle and specificity. We extend the fact-first principle to be the deductive-consequence-first principle. The extended principle expresses the view that all deductive consequences of facts (including facts themselves) are as firm as facts in defeating default conclusions. We have taken into account the priority of deductive consequences in Definition 30.

Specificity is a principle often used to settle conflicting information. It is exemplified in the Tweety bird example (example (9) in Section 1). Imagine that we are in a situation in which we do not have directly observed evidence either showing that Tweety flies, or showing that Tweety does not fly. We must infer one way or the other from the five given premises: Penguins are birds; Birds fly; Penguins do not fly; Tweety is a bird; Tweety is a penguin. The consensus seems to be that we should conclude "Tweety does not fly" from this scenario. If one asks why, the answer is probably that "Tweety is a penguin" is more specific than "Tweety is a bird", since penguins are birds, and Tweety should follow penguins' properties more closely than those of birds in case of conflicts. This intuition can be abstracted as a principle stating that the default about the more specific information has a higher priority. Given that "Penguins are birds", "Penguins do not fly" is more preferable than "Birds fly".

The priority rank among defaults is expected to be understood in the context of breaking conflicts. Without the proper context, sometimes it does not make much sense to claim one default has a higher priority than another. For instance, given that "Penguins are birds", one may question in which sense "Penguins are funny animals" should have a higher priority than "Birds fly"? We need to come up with a story to get an appropriate context against which the comparison of the priority between two defaults will start to make sense. Suppose that we are given two more premises: "Funny animals are fat"; "Flying animals are not fat". Is Tweety, as both a penguin and a bird, fat or not? It seems that we would like to conclude that Tweety is fat. That is because in this particular context, we give a higher priority to "Penguins are funny animals" than to "Birds fly".

The preference order among defaults that is introduced by the specificity principle is determined by the specificity of antecedents of defaults to be ordered. If "Penguins are birds" were not known, "Penguins do not fly" and "Birds fly" would not be comparable on preferability. The formal characterization of the specificity principle that we give below will and must refer to the specificity of antecedents, while intuition may suggest that it is not necessary in some circumstances. For example, even if "Penguins are birds" is not explicitly stated as a premise, as in example (8) in Section 1, some people may feel unable to resist following "Penguins do not fly" and reaching the conclusion that "Tweety does not fly." This does not exhibit that the prerequisite of the specificity of antecedents is not necessary. Rather, it exhibits that we let a hidden premise "Penguins are birds" that exists in our background knowledge sneak into the inference. To exempt this kind of hidden premise, let us consider a modified example. We write down all premises that we are given: "Totos do not fly"; "Tutus fly"; "Tweety is a toto"; "Tweety is a tutu". In this case, our background knowledge contains neither "totos are tutus" nor "tutus are totos". Without an assertion on the specificity between totos and tutus, we truly do not know what to infer. This example has the pattern of the Nixon Diamond. It illustrates that information about the specificity of antecedents is necessary in order to determine the preference order among defaults. The function of the specificity criterion is to break conflicts.

There are other relevant principles besides fact-first and specificity. The authority level of the information source may affect the trustworthiness of the information. Conflicting information from different sources could have different weights in inferences. Here is an example of this type: The company payroll office told John that his salary was deposited to his bank account. The bank officer told John that they did not have the deposit record of his salary. Was the money in John's account? John would probably believe that the money was not in his account and he had to find out what happened to that money. On warning of some natural disasters like earthquakes or tornadoes, the government often urges people to evacuate their residencies. The decision reflects the tendency to be over protective than under protective. In situations like launching a space shuttle, where the price for even a very slim chance of failure is extremely high, system design decisions must favor over estimating the difficulties and dangers of the task. Legal documents explicitly state which set of rules and regulations take precedence in case of conflicts. All these examples suggest that information is not treated equally. The laws that guide people's preferences for information are empirical. They are quite different from logical laws. They are used as criteria for resolving conflicts between inferences, but not as inference rules that can actually infer something. We tend to believe that this level of principles is not suited to be axiomatized into logic systems. Instead of encoding them into our logic system as an axiom or a rule of inference, we encode them into the preference relations over premises in the second phase of our two-phase framework. The separation of this level of principles from the underlying logic makes the core logic more stable. New principles to resolve conflicts can simply be adopted in the second phase. Changes will not ripple down to the underlying logic system.

The principles discussed above will shape $\succeq$ to be an appropriate relation so that no more and no less than our intuitively-expected conclusions will emerge. To exemplify our second phase, we encode the most salient specificity into the binary relation $\succeq$ on $\mathsf{CCL}_M(\Gamma)$, and label it as specificity preferable relation $\succeq_s$. However, this is not meant to be the only principle that guides our preference over premises. Here is the formal definition of $\succeq_s$.

Definition 31. *Given a set Γ of formulas, a specificity preference relation $\succeq_s$over $\mathsf{CCL}_M(\Gamma)$ is the smallest transitive closure satisfying: if $\alpha > \beta, \alpha > \gamma, \beta > \delta \in \mathsf{CCL}_M(\Gamma)$, then $(\alpha > \gamma) \succeq_s (\beta > \delta)$.*

The above definition of $\succeq_s$ ensures that the default with the more specific antecedent takes precedence. The requirement that $\succeq_s$ must be closed under transitivity is to treat the nested Penguin Principle.

5.3 Examples Re-visited

In this section, we re-visit the examples that were set forth in Section 1. Now we can provide formal analysis of these examples based on the notion of nonmonotonic inference relation that has been precisely defined. Given a finite set Γ of formulas and a formula α, $\Gamma \mid\!\sim \alpha$ abbreviates $\Gamma \mid\!\sim_{DC,\succeq_s} \alpha$ in the following justifications to simplify the notation. Also, except (4), in the examples below, $\mathsf{CCL}_M(\Gamma) = \Gamma$. Hence, the relation $\succeq_s$ on $\mathsf{CCL}_M(\Gamma)$ is the same as if it is on Γ.

(1) If it rains, then the ground gets wet. It rains. / The ground gets wet.

(2) Birds fly. Tweety is a bird. / Tweety flies.

The premise sets of examples (1) and (2) can be formalized as $\Gamma_{12} = \{p, p > q\}$. Let $\Delta = \Gamma_{12}$, and thus $q \in C_N(\Delta)$. There is no $\Theta \subseteq \Gamma_{12}$ such that $\neg q \in C_N(\Theta)$. Therefore, $\Gamma_{12} \mid\!\sim q$.

(3) Birds fly. Tweety is a bird. Tweety does not fly. / Tweety does not fly.

$\Gamma_3 = \{\neg q, p, p > q\}$ represents the premise set of example (3). $\neg q \in C_M(\Gamma_3)$, so $\Gamma_3 \mid\!\sim \neg q$. Since, $C_M(\Gamma_3)$ is consistent, according to the Theorem 10, $\Gamma_3 \not\mid\!\sim q$.

(4) Whales are mammals. Marine creatures normally are not mammals. Willy is a whale. Willy is a marine creature. / Willy is a mammal.

$\Gamma_4 = \{p \rightarrow q, r > \neg q, p, r\}$ is the premise set of example (4). $q \in C_M(\Gamma_4)$. Therefore, $\Gamma_4 \mid\!\sim q$. Since, $C_M(\Gamma_3)$ is consistent, by 10, $\Gamma_4 \not\mid\!\sim \neg q$.

The analysis correctly predicts that the deductively derivable conclusion "Willy is a mammal" is retained to be the global conclusion. In this example, $\mathsf{CCL}_M(\Gamma_4) = \{p \rightarrow q, r > \neg q, p, r, q\} \neq \Gamma_4$, and $\succeq_s$ on $\mathsf{CCL}_M(\Gamma_4) \times \mathsf{CCL}_M(\Gamma_4)$ is $\varnothing$. Though the preference relation $\succeq_s$ on $\mathsf{CCL}_M(\Gamma_4)$ is not used to justify $\Gamma_4 \mid\!\sim q$ here, it shows that sometimes $\mathsf{CCL}_M(\Gamma)$ could be different from Γ.

(5) If it rains, the ground gets wet. It rains and the wind blows. / The ground gets wet.

(6) Quakers are pacifists. Republicans are not pacifists. Nixon is a Quaker. / Nixon is a pacifist.

Examples (5) and (6) demonstrate that the additional irrelevant information does not hurt the previously drawn conclusion. The minor difference between examples (5) and (6) is that the additional information in example (5) is a fact, while in (6), it is a default. The premise set Γ_5 is $\{r, p, p > q\}$ and Γ_6 is $\{r > s, p, p > q\}$. Let $\Delta = \{p, p > q\}$. $q \in C_N(\Delta)$. There is no $\Theta \subseteq \Gamma_5$ (or Γ_6) such that $\neg q \in C_N(\Theta)$. Therefore, $\Gamma_5 \mid\!\sim q$ (or $\Gamma_6 \mid\!\sim q$).

(7) Quakers are pacifists. Republicans are not pacifists. Nixon is a Quaker. Nixon is a republican. / Nixon is a pacifist?? Nixon is not a pacifist??

(8) Birds fly. Penguins do not fly. Tweety is a bird. Tweety is a penguin. / Tweety flies?? Tweety does not fly??

Examples (7) and (8) reflect the nonmonotonic inference pattern known as the *Nixon Diamond.* Their premise set Γ_{78} is $\{p, q, p > \neg r, q > r\}$, and $\succeq_s$ on $\Gamma_{78} \times \Gamma_{78}$ is $\varnothing$. Let $\Delta = \{p, p > \neg r\}$ and $\Theta = \{q, q > r\}$. We have $\neg r \in C_N(\Delta)$ and $r \in C_N(\Theta)$. Since $\succeq_s = \varnothing$, neither $\Delta \rhd \Theta$ nor $\Theta \rhd \Delta$. Then, $\Gamma_{78} \not\mid\!\sim \neg r$, and $\Gamma_{78} \not\mid\!\sim r$, either.

(9) Penguins are birds. Birds fly. Penguins do not fly. Tweety is a bird. Tweety is a penguin. / Tweety does not fly.

This example is the famous *Penguin Principle.* It shows that the common tendency to favor more specific information tips the balance of the tie observed in the Nixon Diamond. Consequently, the default conclusion from the more specific default sustains to be the global conclusion. The premise set Γ_9 of example (9) is $\{p, q, p > q, p > \neg r, q > r\}$. $\succeq_s$ on $\Gamma_9 \times \Gamma_9$ is $\{\langle p > \neg r, q > r\rangle\}$. Let $\Delta = \{p, p > \neg r\}$. Thus, $\neg r \in C_N(\Delta)$. Even though there is $\Theta = \{q, q > r\}$ such that $r \in C_N(\Theta)$, $\Delta \rhd \Theta$ because $\langle p > \neg r, q > r\rangle$ is in $\succeq_s$. As a matter of fact, for any $\Theta \subseteq \Gamma_9$ such that $r \in C_N(\Theta)$, Θ must contain $q > r$ and hence $\Delta \rhd \Theta$. Therefore, $\Gamma_9 \mid\!\sim \neg r$ but not $\Gamma_9 \mid\!\sim r$.

The proof can be applied to the extended Penguin Principle with n iterations, and the result holds. Let the premise set Γ_9' be $\{p_i, p_i > p_{i+1}, p_{2k-1} > \neg r, p_{2k} > r \mid 1 \leqslant i, k \leqslant n\}$. $\succeq_s$on $\Gamma_9' \times \Gamma_9'$ is the transitive closure of $\{\langle p_{2k-1} > \neg r, p_{2k} > r\rangle, \langle p_{2k-1} > \neg r, p_{2k+1} > \neg r\rangle, \langle p_{2k} > r, p_{2k+1} > \neg r\rangle, \langle p_{2k} > r, p_{2k+2} > r\rangle \mid 1 \leqslant k \leqslant n\}$. This is to say that $\succeq_s = \{\langle p_{2k-1} > \neg r,\ p_{2j} > r\rangle, \langle p_{2k-1} > \neg r,\ p_{2j-1} > \neg r\rangle, \langle p_{2k} > r,\ p_{2j+1} > \neg r\rangle, \langle p_{2k} > r,\ p_{2j} > r\rangle \mid 1 \leqslant k, j \leqslant n, \text{ and } k < j\}$. Let $\Delta = \{p_1, p_1 > \neg r\}$. Thus, $\neg r \in C_N(\Delta)$. For any $\Theta \subseteq \Gamma_9'$ such that $r \in C_N(\Theta)$, Θ must contain some $p_m > r$ where $1 < m$. $\Delta \rhd \Theta$ because $\langle p_1 > \neg r, p_m > r\rangle \in \succeq_s$ and also $\langle p_m > r, p_1 > \neg r\rangle \notin \succeq_s$. Thus, $SPS_\Gamma(\neg r) \gtrdot SPS_\Gamma(r)$. Therefore, $\Gamma_{9'} \mid\!\sim \neg r$ but not $\Gamma_{9'} \mid\!\sim r$.

Our approach of "going forth and drawing back" demonstrates its advantage of managing the nested Penguin Principle with ease, regardless of its complexity.

(10) College students are adults. Adults can drive. John is a college student. / John can drive.

(11) College students are adults. Adults are employed. John is a college student. / John is employed ??

Example (10) illustrates the successful use of pointwise transitivity, while example (11) shows a failed application if it is used outside the restricted condition. There is an implicitly used premise in the form of $((\alpha \wedge \beta) > \gamma)$ that causes the difference between success and failure. The actual premise set Γ_{10} used in these two examples is $\{(p \wedge q) > r, p, p > q, q > r\}$. $\succeq_s$ on $\Gamma_{10} \times \Gamma_{10}$ is $\varnothing$. The premise $((\alpha \wedge \beta) > \gamma)$ is true in example (10), but not true in example (11). Let $\Delta = \Gamma_{10}$. According to $\mathsf{Th}_{DC}4$, $r \in C_N(\Delta)$. There is no $\Theta \subseteq \Gamma_{10}$ such that $\neg r \in C_N(\Theta)$. Therefore, $\Gamma_{10} \mid\!\sim r$.

In addition, we can verify that our framework supports the *Double Diamond* defeasible inference pattern that generates floating conclusions. In particular, we will be able to nonmonotonically infer that Nixon is politically extreme from the premises: Nixon is both a Quaker and a Republican. Nixon is likely a dove if he is a Quaker, a hawk if he is a Republican. Nixon is likely to be politically extreme if he is either a dove or a hawk. If Nixon is a dove then he is not a hawk, and vice versa. The formal representation of the premise set is $\Gamma = \{q, r, q > d, r > h, (d \vee h) > e, d \to \neg h, h \to \neg d\}$. $\mathsf{CCL}_M(\Gamma) = \Gamma$. $\succeq_s$on $\Gamma \times \Gamma$ is $\varnothing$. There is a subset $\Delta = \{q, q > d, (d \vee h) > e\}$ of Γ such that $e \in C_N(\Delta)$. There is no $\Theta \subseteq \Gamma$ such that $\neg e \in C_N(\Theta)$. Therefore, $\Gamma \mid\!\sim e$.

We have shown that system DC together with $\succeq_s$ are sufficient to successfully account for all the benchmark examples listed in Section 1. The advantages of endorsing DMP are appealing. The deployment of DMP turns the inference of default conclusions from an appromixation of monotonic inference using *Modus Ponens* to an independent type of inference. Establishing a proof of a certain default conclusion does not involve any stepwise consistency check. Reasoners can focus more on what to infer, rather than how to infer (i.e., license the inference to go forward at *every* step based on the current consistency status). The consistency check needs to be done only *once* at the very end when a default conclusion is about to be accepted as a global conclusion.

We advocate our two-phase approach to treat defeasible reasoning, as this approach matches with the basic features of the type of inference that we have summarized. The results that we presented in this paper are based on the minimal setup of the two-phase framework under which benchmark examples are tested. Future extensions could be done in two directions.

One direction is to strengthen the underlying logic by adding more axioms of the "normally follow" operator $>$. For example, we may add an Antecedent Disjunction (AD) axiom $((\alpha > \gamma) \wedge (\beta > \gamma)) \rightarrow ((\alpha \vee \beta) > \gamma)$. Another extension direction is to enrich the preference relation by adopting more principles to shape the ordering of premises. There are other empirical principles that influence information preference. If we would like to encode the preference of the information from a higher authority, our framework can be properly extended to reflect this principle.

6 Conclusions

We have taken the modal conditional approach to formalize defaults, and investigated the kind of inference involving defaults. We observed several fundamental features of such defeasible inference. Roughly speaking, a defeasible inference takes the entire premise set into account and warrants those conclusions that are most preferable in the face of conflicts. Conclusions that can be obtained from some premises are considered locally-inferred default conclusions. The final global conclusions that can be warranted from all of the given premises are selected from default conclusions according to some criteria rooted in the preference relation among premises. Based on this analysis, we intended to capture the defeasible feature of such inference in a two-phase structure. First is the "going forth" phase, in which all candidate conclusions can be deduced from a formal system. Second is the "drawing back" phase, in which conflicting candidates are resolved by giving up less preferable ones and retaining the rest.

A logic system DC has been developed to power the local inference. Just as in the classical logic, the process of deriving conclusions for given premises in DC is monotonic. The management of conflicting default conclusions that can be obtained from DC is left for the next stage. The system DC has Default *Modus Ponens* as an axiom to detach the consequent of the "normally follow" connective. The capability of detachment is enhanced by the restricted pointwise transitivity axiom. The system DC is interpreted in set selection function semantics, which is a variant of selection function semantics. It is proved that system DC is sound and complete with respect to the corresponding semantics.

With a formal logic system providing a common ground to draw default conclusions, determining which default conclusions can sustain to be global conclusions for a given premise set means determining which subsets of premises are preferable. A general discussion has been given on how to construct a preference relation on sets of formulas from a preference relation on formulas that is initially given. This step bridges the gap between the commonly referred preference relation among information (like the specificity principle) and a preference relation on sets of premises that we need in the second phase of the framework. The nonmonotonic inference relation is defined with respect to an underlying logic system and a preference ordering for resolving conflicting default conclusions. As the less preferable conflicting conclusions will be filtered out during the drawing back phase, we have shown that the global conclusion set is consistent. Under our approach, the consistency check is done only once in the drawing back phase to determine if there are conflicts. Comparison of the priority of premise sets happens in a finite domain, so the process of choosing between conflicting default conclusions, if needed, is indeed manageable. The system DC and a preference relation $\succeq_s$ coded from specificity are used to exemplify the use of the two-phase structure to justify the benchmark problems.

Unlike traditional characterization of nonmonotonic inference in which the inference step is mingled with the consistency check, we propose a clear separation into two phases. The underlying logic system is orthogonal to the top level filtering mechanism built from preference relation. They can be configured and extended independently. Principles regarding how to detach default conclusions from defaults can be expressed in the underlying logic as its axioms or rules of inference. The general tendency to overweigh a specific group of information can be coded into the preference relation. Putting two phases together can capture the notion of nonmonotonic inference we intend to address. The system DC and the preference relation $\succeq_s$ we defined are particular instances, the combination of which is proved to be sufficient to account for a list of benchmark examples. The two-phase structure we proposed is a general framework in which stronger systems may be developed and more restrictive relations may be defined. Consequently, a larger set of examples could be accounted for, if desired.

Last but not least, the position of two-layers inference presented in this paper also has connections to the recent development of argument systems (e.g., [10, 5, 22, 6, 30, 11]) that formalize the construction, comparison and evaluation of arguments for and against certain conclusions. We have a similar view to theirs that there are competing arguments that lead to conflicting default conclusions, as opposed to getting everything right in the first place in only one argument for a final expected conclusion.

Compared to existing argument systems, our intentional approach has the following properties.

(1) The logical system DC not only provides a natural formalism to represent

defaults, but also has a formal semantics, based on which the soundness and completeness of the logical system could be proved. Such property does not belong to existing argument systems. On the one hand, for a rule-based argument system (e.g., [13, 11]), although defaults could be naturally represented by defeasible rules, there is no formal semantics. As a result, the completeness of the system could not be verified. On the other hand, for a system based on first-order logic [12], to express a default by a material implication is not a natural way.

(2) As all arguments for default conclusions share the same inference engine provided by the system DC, the comparison of the strength of arguments is reduced to the comparison of the strength of alternative premise sets that support alternative default conclusions. Thus, the problems caused by zombie paths [18] do not apply to our methodology.

References

[1] Alchourrón C. E. and Makinson, D. Hierarchies of regulations and their logic. In R. Hilpinen (Ed.), *New Studies in Deontic Logic*, pages 125–148, D. Reidel, Dordrecht, 1981.

[2] Asher, N., and Mao, Y. Negated defaults in commonsense entailment. *Bulletin of the Section of Logic*, 30(1):41–60, 2001.

[3] Asher, N. and Morreau, M. Commonsense entailment: A modal theory of nonmonotonic reasoning. In J. Mylopoulos and R. Reiter (Eds), *Proceedings of the Twelfth International Joint Conference on Artificial Intelligence*, pages 387–392, Morgan Kauffman, Los Altos, California, 1991.

[4] Anderson, A.R., Belnap, N.D., and Dunn, J.M. *Entailment: the logic of relevance and necessity, vol II.* Princeton University Press, 1992.

[5] Baroni, P., Caminada, M., and Giacomin, M. An introduction to argumentation semantics, 26. *The Knowledge Engineering Review* 26:365–410, 2011.

[6] Besnard, P., and Hunter, A. Constructing argument graphs with deductive arguments: a tutorial. *Argument & Computation*, 5:5–30, 2014.

[7] Boutilier, C. Conditional logics of normality: A modal approach. *Artificial Intelligence*, 68:87–154, 1994.

[8] Delgrande, J. A First-Order Conditional Logic for Prototypical Properties. *Artificial Intelligence*, 33:105–130, 1987.

[9] Delgrande, J. An Approach to Default Reasoning Based on a First-Order Conditional Logic. (Revised report) *Artificial Intelligence*, 36:63–90, 1988.

[10] Dung, P.M. On the acceptability of arguments and its fundamental role in nonmonotonic reasoning, logic programming and n-person games. *Artificial Intelligence*, 77:321–357, 1995.

[11] García, A.J. and Simari G.R. Defeasible logic programming: DeLP-servers, contextual queries, and explanations for answers. *Argument & Computation*, 5:63–88, 2014.

[12] Gorogiannis, N. and Hunter H. Instantiating Abstract Argumentation with Classical Logic Arguments: Postulates and Properties. *Artificial Intelligence*, 175:1479–1497, 2011.

[13] Governatori, G., Maher, M.J., Antoniou G., and Billington, D. Argumentation Semantics for Defeasible Logic. *Journal of Logic and Computation*, 14:675–702, 2004.

[14] Horty, J. F., Thomason, R. H., and Touretzky, D. S. A skeptical theory of inheritance in nonmonotonic semantic networks. *Artificial Intelligence*, 42:311–348, 1990.

[15] Horty, J. F. Skepticism and floating conclusions. *Artificial Intelligence*, 135:55 - 72, 2002.

[16] Kraus, S., Lehmann, D., and Magidor, M. Nonmonotonic reasoning, preferential models, and cumulative logics. *Artificial Intelligence*, 44:167–207, 1990.

[17] Lifschitz, V. On the satisfiability of circumscription. *Artificial Intelligence*, 28:17–27, 1986.

[18] Makinson, D., and Schlechta, K. Floating conclusions and zombie paths: two deep difficulties in the 'directly sceptical' approach to inheritance nets. *Artificial Intelligence*, 48:199–209, 1991.

[19] Mao, Y. A Formalism for Nonmonotonic Reasoning Encoded Generics. Dissertation, University of Texas at Austin, 2003.

[20] Mao, Y., and Zhou, B.(2007, January) *The Cause and Treatments of Floating Conclusions and Zombie Paths.* (Paper presented at The Seventh IJCAI International Workshop on Nonmontonic Reasoning, Action and Change, India)

[21] McCarthy, J. Circumscription—a form of non-monotonic reasoning. *Artificial Intelligence*, 13:27–39, 1980.

[22] Modgil, S., and Prakken, H. A general account of argumentation with preferences. *Artificial Intelligence*, 195:361–397, 2013.

[23] Morreau, M. Allowed arguments. In J. Mylopoulos and R. Reith (Eds), *Proceedings of the Sixteenth International Joint Conference on Artificial Intelligence*, pages 1466–1472. Morgan Kaufmann, Los Altos, California, 1995.

[24] Nute, D. Conditional Logic. In D.M. Gabbay and F. Guenthner (Eds), *Handbook of Philosophical Logic, VII*, pages 387–439, Kluwer Academic Publishers, Dordrecht, 1994.

[25] Nute, D. *Topics in Conditinal Logic.* Reidel, Dordrecht, 1980.

[26] Pelletier, F. J., and Asher, N. Generics and defaults. In J. van Benthem and A. ter Meulen (Eds), *Handbook of Logic and Language*, pages 1125–1177. The MIT Press, Cambridge, MA, 1997.

[27] Prakken, H. Intuitions and the modelling of defeasible reasoning: some case studies. *Proceedings of the Ninth International Workshop on Non-monotonic Reasoning*,

Toulouse, pages 91–99, 2002.

[28] Reiter, R. A logic for default reasoning. *Artificial Intelligence*, 13:81–132, 1980.

[29] Shoham, Y. *Reasoning about Change*. The MIT Press, Cambridge, Massachusetts, 1988.

[30] Toni, F. A tutorial on assumption-based argumentation. *Argument & Computation*, 5:89–117, 2014.

Received May 2015

www.ingramcontent.com/pod-product-compliance
Lightning Source LLC
Chambersburg PA
CBHW081013040426
42444CB00014B/3187